Walter Voigt

Fachenglisch für
Chemisch-technische Assistenten

Ebenfalls in der CTA-Reihe sind erschienen:

V. Joos
Physik für Chemisch-technische Assistenten
1984, ISBN 3-527-30853-9

F. Bergler
Physikalische Chemie für Chemisch-technische Assistenten
2. Auflage, 1991, ISBN 3-527-30846-6

Walter Voigt

Fachenglisch für Chemisch-technische Assistenten

2., durchgesehene Auflage

Herausgegeben von W. Fresenius, B. Fresenius,
W. Dilger und W. Flad

Verantwortlicher Herausgeber für diesen Band:
W. Flad

 WILEY-VCH

Walter Voigt, OStR
Chemisches Institut Dr. Flad
Breitscheidstraße 127
70176 Stuttgart

Titelbild: Chemisches Institut Dr. Flad, Stuttgart
         M. Müller, Weinheim

1. Auflage 1983
2. Auflage 1988

Die Deutsche Bibliothek – CIP-Einheitsaufnahme

Ein Titeldatensatz für diese Publikation ist bei Der Deutschen Bibliothek erhältlich

ISBN: 978-3-527-30876-7

Gedruckt auf säurefreiem Papier

© WILEY-VCH Verlag GmbH, Weinheim (Federal Republic of Germany), 2002

Geleitwort zur 2. Auflage

Für eine erfolgreiche Arbeit von Schülern und Lehrern sind brauchbare Lehr- und Lernmittel unverzichtbar. Gerade für die berufsbildenden Schulen aber fehlen oft die geeigneten Schulbücher. Speziell die Berufsfachschulen und Berufskollegs für Chemisch-technische Assistenten müssen sich oft behelfen und haben teilweise schon zur Selbsthilfe gegriffen und eigene Lehrmittel verfaßt. Doch hierbei muß ein hoher Aufwand betrieben werden und viel Wissen und Erfahrung bleiben dennoch nur wenig genutzt. Ziel dieser Buchreihe ist es nun, die jahrelange Ausbildungserfahrung vieler Kollegen allen interessierten Schülern und Dozenten zur Verfügung zu stellen.

Besondere Schwierigkeiten bereitet naturgemäß an einer Chemieschule der vorgeschriebene Englischunterricht, der sich selbstverständlich an der späteren Berufstätigkeit orientieren muß. Doch meist verfügt der Chemiker nur über Englischkenntnisse für den Eigengebrauch und der Kollege für das Fach Englisch nicht über ausreichendes Chemiewissen. Vor allem aber mangelt es an einem speziellen Schulbuch für dieses Fachenglisch. Es soll und darf kein Chemiebuch in englischer Sprache sein. Vielmehr sollen interessante und ansprechende Berichte über die verschiedenen Bereiche der Chemie dem Schüler unversehens den erforderlichen Fachwortschatz und Kenntnisse der Grammatik vermitteln. Wenn gut gemacht, muß dieses Buch sowohl den Chemiker als auch den Anglisten beim Lesen fesseln und dem Schüler eine willkommene Abwechslung bieten.

Ich bin überzeugt, daß dem Autor W. Voigt ein solches Buch gelungen ist. Er unterrichtet seit vielen Jahren in Stuttgart und hat sich mit viel Engagement und ebensoviel Erfolg nach seinem Sprachstudium auch noch in die Chemie eingearbeitet. Die glückliche Verbindung beider Fachgebiete wird bei der Arbeit mit diesem Buch offenbar. Ich wünsche ihm eine große Verbreitung und den Benutzern viel Spaß und Erfolg.

Daß inzwischen die 1. Auflage bereits vergriffen ist, macht deutlich, daß nicht nur Chemieschüler in Deutschland dieses Buch gerne benutzen. Der zwar korrigierten aber sonst unveränderten 2. Auflage wünsche ich weiterhin eine erfolgreiche Verwendung.

Stuttgart, im Februar 1988   Für die Herausgeber

Wolfgang Flad

Vorwort zur 2. Auflage

Das vorliegende Buch ist in erster Linie als Unterrichtswerk für Schüler gedacht, die an Berufsfachschulen (Berufskollegs) eine Ausbildung zum Chemisch-technischen Assistenten erfahren. Darüber hinaus bietet es für Schüler der gymnasialen Oberstufe, Chemiestudenten der ersten Semester und für alle an der Chemie Interessierten einen breitgefächerten Einblick in diese Wissenschaft und einen Einstieg in englische Fachtexte. Dabei wurde die Themenauswahl so getroffen, daß verschiedene Bereiche der Chemie behandelt werden, die ebenso abwechslungsreich wie interessant sind:

> 'Chemie – Im Wandel der Zeiten',
> 'Chemie – In Theorie und Praxis',
> 'Chemie – Im Alltagsleben' und
> 'Chemie – Auf den zweiten Blick'.

Jeder dieser Themenbereiche ist unterteilt in sogenannte 'Units', die aus einem Text, dem dazugehörenden Vokabelteil, Fragen zum Text, Paraphrasierungen einzelner Begriffe, Grammatik und sich anschließenden Übungen bestehen.

Das erste Kapitel 'Chemistry – Through the Ages' kann verständlicherweise nur schlaglichtartig einige wesentliche Stationen und Entwicklungen aufzeigen, die diese Wissenschaft seit ihren Anfängen bestimmt haben. Durch gezielte Erweiterungen und Zusatzinformationen seitens des Unterrichtenden oder durch Referate von Schülern zu bestimmten Teilgebieten kann eine gründlichere Information erzielt werden.

Eine breitangelegte Auswahl an Texten im Kapitel 'Chemistry – In the Laboratory' ermöglicht ein Sichvertrautmachen mit chemischen Fachausdrücken der englischen Sprache. Dieser Teil ist auch konzipiert als Vorbereitungshilfe für die seit einigen Jahren existierende internationale Berufsabschlußprüfung, die von der International Schools Association (ISA) mit Sitz in Genf abgenommen wird.

Im dritten Teil 'Chemistry – In Everyday Life' werden Themen berührt, die heutzutage in aller Munde sind und Stoff für kontroverse Diskussionen bieten: 'Saurer Regen', 'Kunstdünger', 'Pestizide', 'Wasserverschmutzung' usw. Insbesondere hier kann und sollte durch zusätzliche Textangebote und Hinweise auf die Gefahren eines unkontrollierten Einsatzes chemischer Erzeugnisse, aber auch auf die Unerläßlichkeit einer sinnvoll angewandten Chemie aufmerksam gemacht werden.

Im letzten Kapitel 'Chemistry – At Second Sight' werden Bereiche unseres Lebens vorgestellt, die auf den ersten Blick anscheinend nicht allzuviel mit Chemie zu tun haben, aber bei genauerem Betrachten stark von ihr beeinflußt werden oder ohne deren Erkenntnisse und Möglichkeiten nicht mehr denkbar wären wie z. B. die Kosmetik, der Sport, die Kriminalistik. Da auch dies eine willkürliche Stoff- und Textauswahl darstellt, kann eine Erweiterung der Themenpalette motivierend und interessant sein.

Wie schon erwähnt, schließt sich an jeden Text ein Vokabelteil an, der die unbekannten und schwierigen Wörter erklärt.

Die Anzahl der folgenden 'Questions' ist bewußt geringgehalten, um ein 'Erschlagen' des Schülers durch einen übergroßen Fragenblock zu vermeiden. Sie lassen sich, wie auch die Paraphrasierungen, gut als schriftliche Hausaufgabe verwenden, während weitere Verständnis- und Inhaltsfragen sich aus dem konkreten Unterrichtsablauf entwickeln sollten.

Aus Gründen der für den Unterricht zur Verfügung stehenden Zeit kann und will das Buch nicht den Anspruch eines lückenlosen grammatikalischen Nachschlagewerks erheben, sondern beinhaltet nur die wichtigen Gebiete.

Der Leitgedanke war, den Schüler durch Vorgabe eindeutiger Beispiele dazu zu bringen, die entsprechenden Regeln selbst abzuleiten, zu formulieren und niederzuschreiben. Weitergehende Grammatikprobleme sollten durch den Unterrichtenden angemessen angegangen und geklärt werden. In den sich anschließenden Übungen können die erworbenen Regelkenntnisse dann umgesetzt und angewendet werden.

Schließlich möchte ich an dieser Stelle meinem Herausgeber Wolfgang Flad für dessen rat- und tatkräftige Unterstützung danken, ebenso wie den Kollegen, die durch ihre Anregungen und Hilfestellungen zu einem hoffentlich nützlichen Buch beitrugen.

Die Konzeption des Buches wurde auch für die 2. Auflage unverändert beibehalten. Viele Zuschriften haben mich in dieser Absicht bestätigt. Danken möchte ich allen Kolleginnen und Kollegen, die mich auf Schreib- oder Druckfehler hingewiesen und so zur Verbesserung des Buches beigetragen haben. Dabei habe ich erfreut feststellen dürfen, daß dieses Buch auch vielfach außerhalb der deutschen Chemieschulen Verwendung findet.

Stuttgart, im Februar 1988   *Walter Voigt*

Inhaltsverzeichnis

1 Chemistry – Through the Ages

Beginnings of Chemistry
Alchemy
Iatrochemistry – Its Rise and Fall
Chemistry in the 19th and 20th Century

Beginnings of Chemistry

Chemistry as an art goes back to the most ancient civilizations. Sumerians, Egyptians and the ancient inhabitants of India and China probably knew about different processes, though we don't know exactly how they carried them out. The Egyptians were rather skilful in the application of dyes and knew how to use mordants in dyeing. They melted ores and made weapons, armours, ornamental objects and tools. They prepared medicines from herbs, animal and mineral substances, leather by a process of tanning, and beer and wine by fermentation. Moreover, they were capable of producing glass and also knew how to vary its colour by adding suitable substances. They used copper, gold, silver and **electron**, the alloy of the two metals, to fashion objects from these substances.

The pursuit of experimental chemistry in Egyptian times appears to have been in the hands of priests. Records exist that contain precise and clear instructions for carrying out chemical and other operations, such as

– making artificial gems,
– dyeing,
– bleaching discoloured pearls, and
– preparing amalgams and alloys.

It was not until the 1st century that the Alexandrians really considered and even attempted to convert a base metal into gold, which would later become the obsession of thousands of alchemists.

One can also suppose that the value of experiment as an aid to discovery was recognized during an early period of Egyptian civilization. We can attribute the process of embalming to design rather than accident.

Man has always wondered about the origin and constitution of matter, and how it reacts under different conditions. But it was only in the 6th century BC that attempts were made at an explanation.

Greek philosophers (Thales, Anaximander, Anaximenes) believed that

all things were made from one indestructible and primordial substance, differing only in their conception of this material.

The first of these believed that **water** was the fundamental substance, Anaximander suggested a hypothetical **principle** and the latter considered **air** as the primary matter. Later, fire was held to be the original element (Heraclitus of Ephesus).

In the 5th century BC these theories were expanded by Empedocles. Now, matter in all its forms consisted of four unchangeable elements:

| earth | water | air | fire |

Democritus considered matter as being composed of atoms, being indestructible and ever moving in an otherwise vacuous space. It was believed that these atoms consisted of only one substance and existed in different sizes and types; that combinations and separations were responsible for changes and all the other phenomena associated with material things.

Plato and his pupil Aristotle disapproved of the idea that there really could exist such a vacuum. For them, matter had to be

infinitely divisible, continuous and
completely filling space.

So, they fell back upon the four-element theory of Empedocles and rejected the atomic theory of Democritus.

In addition Aristotle introduced a 5th element, the unchanging **ether**, of which the stars and the celestial sphere were supposed to be composed.

According to him

heat   **coldness**   **wetness**   and **dryness**

were the only inherent components of matter:

– dryness plus coldness	composing	Earth
– wetness plus coldness	composing	Water
– wetness plus heat	composing	Air
– dryness plus heat	composing	Fire

These elements were not considered to be identical with the real substances of the same names:

they were understood rather as principles; the name suggesting the tangible manifestation of its character: earth = solidity etc.
The hypothesis that these elements could be transformed one into another was based on observations:

– when heat was added to water the latter became gaseous;

– when a combustible solid burned it was changed into fire, etc.

It was a long time before Aristotle's theories were rejected, experiments favoured on a large scale and the atomic theory revived in philosophy.

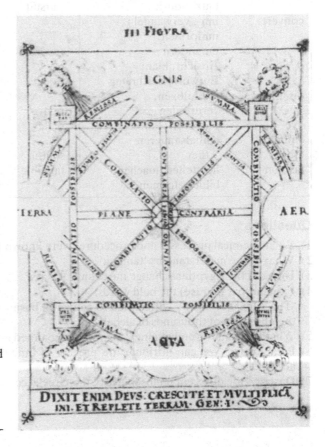

Vocabulary

aid	Hilfe, Unterstützung	favour	bevorzugen
alloy	Metallegierung, Legierung	gaseous	gasförmig
		gem	Edelstein
amalgam	Amalgam, fig. Mischung	herb	Kraut, Heilkraut
		infinite	unendlich, unbegrenzt
ancient	alt, aus alter Zeit; altertümlich	inherent	innewohnend, zugehörend, eigen
application	An-, Verwendung, Gebrauch	manifestation	Offenbarung, (deutliches) Anzeichen, Symptom
armour	Rüstung, Panzer(ung)		
artificial	künstlich	moistness	Feuchte
attempt	versuchen	mordant	Beize; Ätzwasser
attribute	zuschreiben	obsession	Besessenheit, fixe Idee
base	minderwertig, unedel		
bleach	bleichen	ore	Erz
celestial	Himmels . . .	primordial	ursprünglich, uranfänglich, Ur . . .
combustible	brennbar, leicht entzündlich	pursuit	Streben, Trachten, Jagd, Beschäftigung
convert	um-, verwandeln, umformen, umändern	record	Aufzeichnung, Dokument
design	Absicht, Plan; Entwurf, Zeichnung	reject	ab-, zurückweisen, verwerfen
disapprove	mißbilligen, verurteilen	revive	wiederbeleben
dye	Farbstoff, Farbe	scale	Maßstab, Umfang; Waagschale; Skala
embalm	einbalsamieren		
ether	Äther	skilful	geschickt, gewandt
fashion	herstellen, machen, bilden, formen	tangible	greifbar, fühlbar

Questions

1) What chemical processes and procedures were known to ancient civilizations?
2) What do we understand by 'tanning'?
3) How did the Egyptians change the colour of glass?
4) Explain (paraphrase) the bold words:
   'The Alexandrians . . . **attempted** to **convert** a **base** metal into gold . . .' and '. . . the **obsession** of thousands of alchemists'.
5) What is meant by 'embalming'? Who was embalmed and why?
6) Paraphrase: 'We can attribute the process of embalming to design rather than accident'.
7) Describe the different philosophies of the Greeks concerning matter.

Grammar

Passive Voice

Priests **embalmed** the corpse of the Egyptian Pharaoh Tutankhamen.	– A verb that has a direct object is in the **active voice**.
The corpse of the Egyptian Pharaoh Tutankhamen **was embalmed** by priests.	– If the subject is acted upon, the verb is in the **passive voice**. It is mostly used in newspaper reports, scientific, technical and historical descriptions.

The simple passive

to be magnetized	– magnetisiert werden
to be seen	– gesehen werden

Infinitive

active	**passive**
to see	to be seen
to have seen	to have been seen

The powder has been analysed successfully.

Passive denoting a state or action

The cable **is repaired** = (1) . . . ist repariert (state)
(2) . . . wird repariert (action)

To avoid misunderstandings, use the extended forms when expressing actions:
The cable is being repaired = . . . wird gerade . . .

Transformation

A compressed air mechanism moves the plate.

The plate is moved **by** a compressed air mechanism (1).

They carried out the analysis.

The analysis was carried out (2).
Why isn't the by-agent used in the second example? ....

....

Transformation: passive with two objects

Active	We showed **the engineers**	**the test results**
	(indirect object)	(direct object)
Passive	**The engineers** were shown the test results.	
	The test results were shown to the engineers.	

Prepositional object

Active	They sent <u>for</u> the TV technician.
Passive	The TV technician **was sent <u>for</u>**.

Form passive sentences with the following phrasal verbs:

to look after	to hear of	to listen to
to hand over	to dig out	to pull down

Exercises

A   Put into the passive voice:

1) We usually refer to these changes as Verner's law.
2) The scientists were checking all the data.
3) Students must never depart from this principle.
4) They have not thought of this drawback.
5) The committee congratulated the authors on their excellent contribution to the study of the subject.
6) They should give these shortcomings proper attention.
7) People generally suppose that money brings happiness.
8) One cannot eat a banana if nobody has peeled it (2 passives).
9) My parents had already promised me a bicycle for my birthday when they presented me with one as a prize (2 possibilities).
10) One cannot expect a computer to be an electronic psychiatrist.

B   Translate into English. Use passives wherever possible:

1) Die meisten Elektronen sind fest an die einzelnen Atome gebunden.
2) Elektrische Ströme können durch chemische Reaktionen erzeugt werden.
3) Man sagte ihm, daß die Versuche mit großem Erfolg abgeschlossen worden seien.
4) Auf diese Resultate kann man sich nicht verlassen.
5) Das Material wird gerade magnetisiert.
6) Der Verbindungsdraht war entfernt worden.
7) Ein elektrisches Feld wird zwischen den Elektroden erzeugt werden.

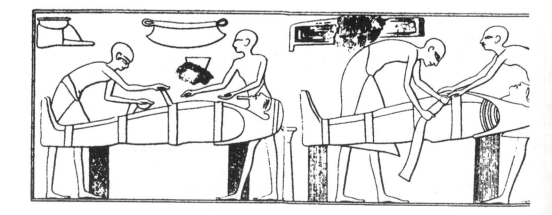

C   Use passive constructions wherever possible:

The Egyptian Pharaoh Tutankhamen is to be Embalmed

Die Leiche (corpse) wird zunächst enthaart (to depilate), dann werden die Eingeweide (entrails) und das Gehirn herausgenommen. Dies geschieht mit wenigen Schnitten und Kunstgriffen (great skill). Die entfernten Organe werden konserviert (to preserve) und in Gefäße (vessel), sog. Kanopen (canopic jars) getan. In die Arterien (artery) wird eine chemische Substanz gespritzt (to inject), und das Innere (inside) des Leichnams wird mit Palmwein (palm wine) ausgespült (to wash out); dann wird der Körper mit einer Mischung aus Wachs (wax), verschiedenen Beeren, Kaneelen (cinnamon), gerösteten Lotuskörnern (roasted lotus seeds) und einer Stoffauspolsterung (padding) gefüllt, alles von Myrre (myrrh) und ätherischen Ölen (essential oil) getränkt (to soak). Nun wird die Leiche 70 Tage in Natron gelegt, welches dem Körper das Wasser völlig entzieht (to dehydrate). Dann wird die Leiche mit Zedernöl (cedar oil) und Balsamen (balm)eingekremt (to rub into) und schließlich mit Leinenbinden (linen cloth) umwickelt (to wrap).

Bei der ganzen Zeremonie wird sehr viel Öl verwendet und das ganze mit Harz (resin) verkleistert (to paste up).

Alchemy

Inside the dark laboratory, the old alchemist moves about, wholly absorbed in his activities.

This time he will succeed – this time there will be no failure – the discovery of the 'philosophers' stone' is close at hand.

Secretly he has worked for years, hidden from the world, secluded from people, as Geber, the Arabian master of alchemistry, had once advised his followers to do: "Do not let anybody, not even your wife, children, or any other person, see what you are doing, because otherwise gold would corrupt the world, for it could then be produced as easily as glass."

At the moment, he is hard-boiling 2000 hens' eggs in enormous pots of boiling water. After having peeled off the shells, he arranges them in a big heap. Gently he heats them until they are totally white. Meanwhile his assistant separates the yolks from the whites, places them into the dung of white horses and lets them rot.

These strange products are distilled over and over again during a period of 8 years for the extraction of a curious red oil and a white liquid.

And now the moment has come to prove that these two powerful solvents can produce the 'philosophers' stone'.

But once again, their 'stone' doesn't change one of the base metals into gold.

Like Bernard Trevisan (1406–1490), innumerable alchemists tried to transmute metals and never succeeded. They crystallized and dissolved all kinds of natural salts and minerals; coagulated and calcined copperas, alum, and every vegetable and animal matter. They worked with dung, flesh, excrement; blood and urine of man. Their operations comprised

> fusion, elevation, evaporation,
> ignition, ascension, descension,
> reverberation, decoction, rectification,
> and innumerable other methods.

They worked for kings and queens, rich people, noblemen, aristocrats and themselves, and many died, after a lifetime of failure, in poverty and distress.

Many of the early European alchemists had been members of religious orders and belonged to the educated classes. Highly intelligent men like Albert Groot, Roger Bacon, Vincent de Beauvais and Raymond Lully had interests that extended beyond the limited range of pure alchemy.

Among later European alchemists there were many of much lower intellect and standing, who corrupted alchemy by fraud and who played upon the credulity and greed of their patrons to get money for their experiments.

The practice of alchemy began to decline in the 16th century. It had, however, a protracted death, for although it no longer attracted the attention of those seriously interested in chemistry after the 17th century, it was not completely forgotten until about the beginning of the 19th century.

As a branch, alchemy contributed nothing of a positive nature to human understanding; as a practical art, although it failed to attain its main objective, it did yield a rich harvest of secondary, unsought information because many important substances, notably the strong mineral acids (nitric, hydrochloric and sulphuric), were discovered by its professors.

Vocabulary

absorbed	vertieft, versunken	descension	Absteigen, Sinken
alum	Alaun	dissolve	(auf-)lösen
ascension	Aufsteigen, Aufstieg	distress	Elend, Not, Qual,
attain	erreichen, erlangen,		Leid
	erzielen	dung	Mist, (Tier)Kot
calcine	(aus-)glühen	elevation	Erhöhung,
coagulate	flockig/klumpig wer-		Emporheben
	den, sich verdicken	evaporation	Verdampfung,
copperas	Vitriol		Verdunstung
corrupt	zugrunde richten,	excrement	Kot
	korrumpieren	extraction	Herausziehen,
credulity	Leichtgläubigkeit		Gewinnung
curious	seltsam, merkwürdig	fraud	Betrug
decoction	Auskochen,	fusion	Verschmelzung
	Absieden	gentle	leicht, sanft

greed	Gier	rectification	Rektifikation
hard-boil	hartkochen	reverberation	Zurückwerfen,
harvest	Ernte, fig.		-strahlen, Widerhall
	Ertrag, Gewinn	rot	(ver)faulen, (-)mo-
heap	Haufen		dern, verwesen
hydrochloric	Salzsäure	secluded	zurückgezogen,
acid			einsam
ignition	An-, Ent-,zünden,	shell	Schale
	Erhitzung, Zündung	standing	Rang, Stellung, Ruf,
nitric acid	Salpetersäure		Ansehen
notable	merklich, beträcht-	sulphuric acid	Schwefelsäure
	lich, bemerkenswert	transmute	umwandeln, -bilden,
objective	Ziel		verwandeln
patron	Gönner, Förderer,	unsought	ungesucht, ungewollt
	Schutz-, Schirmherr	yield	ergeben, (ein-,
peel (off)	(ab)schälen		heraus-)
philosphers'	Stein der Weisen		bringen, abwerfen,
stone			liefern
protract	in die Länge (od.	yolk	Eigelb, Eidotter
	hin)ziehen		
protracted	langwierig;		
	hinhaltend		

Questions

1) What is the 'philosophers' stone?'
2) Explain in your own words how Trevisan attempted to produce it.
3) Name further procedures, processes, experiments, and methods alchemists tried in order to create it.
4) What different materials did they work with?
5) What was the fate of many alchemists?
6) Paraphrase: ' . . . who corrupted alchemy by fraud and who played upon the credulity and greed of their patrons . . .'
7) What were the reasons for the protracted death of alchemy?
8) Were there any positive contributions alchemy made for the further development of chemistry?

Paraphrase:

a) rot   b) greed   c) transmute

Grammar

The Tenses
The present tense (simple and continuous)

Have a look at the following sentences and explain why the present tense is used:

1) Equal volumes of gases at the same temperature and the same very low pressure **contain** the same number of molecules. (Avogadro's Law)

   Simple:

   Like charges **repel,** unlike charges **attract** each other.

   The sun **rises** in the east and **sets** in the west.

   Metals **expand** when heated and **contract** when cooled.   ....

2) The teacher **goes** to his desk and **puts** down his books.
   It **doesn't matter.**   ....

3) **If** he **comes,,** I'll tell him.
   He won't help you in the experiment **unless** you **ask** him.   ....

   My brother **meets** her next Friday.   ....

4) She usually **retires** to the library.   ....

5) I **know** that you come to this institute quite often.   ....

1) Look, the sun **is rising.**
   What **are** you **looking** at?
   Do you understand what I **am saying?**
   His brother **is living** in Berlin.

   Continuous:
   ....

2) English words **are** constantly **changing** their functions.
   I believe your friend **is working** on another book?   ....

3) We **are going** abroad in **July.**
   They **are having** a small party next week.
   How long are you **staying** here?
   When are they **coming** home?   ....

4) I'm afraid it's **getting** late.   ....

Work out the differences in the following examples:

He's **always talking** of how hard up he is.   (. . . aber auch immer . . .)
Other people **are constantly doing**   (. . . aber auch dauernd . . .)
things.

b) Listeners **are always asking** him to
talk about distillation.   (. . . immer wieder . . .)

c) You're **always seeing** something
strange!   (. . . aber auch immer . . .)

   – You know I **always tell** you the
truth.   (. . . immer . . .)
   – You **always say** the right thing.   (. . . immer . . .)
   – He **always keeps** old letters.   (. . . immer . . .)

Exercise

Put the verbs in brackets into the present tense (simple or progressive forms):

1) At present the last argument (to gain) ground.
2) The town (to change) rapidly its appearance.
3) Direct evidence (to lack).
4) It (to freeze) hard.
5) This fact (to provide) at least some information on the point in question.
6) This year he (to work) on a linguistic subject.
7) At present they (to work) in the field.
8) He usually (to come) at half past seven.
9) We (to leave) next week.
10) I (to work) in the lab the whole day.

Iatrochemistry – Its Rise and Fall

It was only during the first half of the 16th century that efforts were made to go beyond the limits of the narrow range of alchemical aims previously set by earlier alchemists. The development owed its origin to the formation of a new school, that of the **iatrochemists,** or physician-chemists, who held as their chief article of faith that the true function of the alchemist was to search for new medicines which would help to cure human ills, and not to seek to turn base metals into gold.

Paracelsus (a Swiss physician surnamed von Hohenheim, 1493–1541), who argued that the human body was fundamentally a chemical system, energetically promoted the new doctrine. In order to correct disorders of the system, manifest as illness, it was necessary to administer chemical remedies. This, however, resulted very often in poisoning patients, as he and his fellow-iatrochemists prescribed dangerous substances, such as preparations of antimony, arsenic and mercury.

aus [6]

The greatest contributions made by Paracelsus to chemistry consisted in

extending its scope, attracting serious investigators,
and to medicine, in
challenging, refuting and finally doing away with ancient dogmas.

But iatrochemists, too, believed in the influence of the stars on human affairs and the power of demons. They were not looking for a gold-making 'philosophers' stone' but a miraculous medicine, the 'elixir of life', providing perfect health and longevity. The new spirit of investigation stimulated interest in chemistry and attracted some of the greatest intellects of the time. As a consequence of this, much was done, especially during the later period, that helped to free both chemistry and medicine from the mysticism and superstition that had previously hindered real progress:

van Helmont (1577–1644) introduced the term 'gas'; de la Boë (1614–1672) observed the relation between combustion and respiration; Tachenius (c. 1620–1690) carried out methods of qualitative analysis and used the term 'salt' as a way of defining the compound of an acid with an alkali; Agricola (1494–1555), known as the 'father of mineralogy'. Although he made contributions to medicine, chemistry, mathematics, theology, and history his most important writings were on mineralogy and mining. Best known of his six books on geological subjects is 'De re metallica'; Glauber (1604–1670); apart from other highly significant findings he became famous for the salt which later bore his name.

The decline of iatrochemistry and the rise of scientific chemistry started, more or less, with the publication of a book entitled 'The Sceptical Chymist'. In it, Boyle (1627–1691) ridiculed the obscurity and mysticism of the alchemists and iatrochemists, criticized their methods and refuted their theories. For him, true knowledge could only be gained by the inductive method, truths being inferred by observations of natural processes and experiments. He especially tried to determine the composition of substances and thus paid considerable attention to methods of chemical analysis. He disapproved of the Aristotelian and alchemistic concept of the elements and argued that only substances, not being composed of 'any other bodies', forming the ultimate ingredients of compound bodies, should be regarded as elements.

Boyle has been called 'Father of Modern Chemistry' by some, and there is no doubt that the new spirit of inquiry which he introduced marked an important turning point in the history of chemistry. But not until 100 years later were the older ideas and methods finally displaced.

Man's attention has always been attracted by the phenomenon of combustion. Two contradictory standpoints were held:

– Becher (1635–1682) postulated the existence of three modified elements, principles or 'earths', as the ultimate constituents of substances and further explained that combustion, in its general sense, including the calcination (oxidation) of metals, accounted for the escape of the combustible principle from the burning body.

– Aware of the fact that metals gain in weight on calcination, Boyle thought that a substance of the nature of fire, having an appreciable weight, was added to a material during combustion.

The development of Becher's theory into a general system of chemistry was due almost entirely to Stahl (1660–1734), a German physician, to whose ardour and energy the doctrine also owed its propagation. Stahl applied, though he did not invent, the name **phlogiston**, from the Greek word, 'phlogistos', meaning 'burnt'; and he endeavoured to explain all known chemical phenomena by reference to this hypothetical entity.

Although there was nothing to confirm the theory (experiments weren't carried out), it existed until the beginning of the last quarter of the 18th century.

It is to be noted that we owe the first useful interpretation of combustion and calcination to the French chemist Lavoisier (1743–1794). By arguments based on known facts, and **by experiment** he proved beyond reasonable doubt that these changes are due to the combustion of the affected substance with **oxygen**, thus completely disproving the phlogiston theory.

Lavoisier's theories did not immediately gain acceptance. It needed the brilliant research of Cavendish which led to the discovery of the quantitative composition of water to complete the evidence.

When Lavoisier was informed of Cavendish's discovery in 1783, he proceeded to verify it and immediately recognized its importance. Interpreting it in accordance with his new theories he directly stated that water was formed by combustion in oxygen of 'inflammable air', which he re-named **hydrogen.** Phlogiston was done away with and chemistry appeared in a completely new light.

Vocabulary

in accordance with	in Übereinstimmung mit	contradictory	entgegengesetzt, widersprechend
account for	erklären	displace	ersetzen, versetzen, -lagern, verdrängen
administer	verabreichen, (ein-)geben; verwalten	disorder	Störung, Erkrankung
affect	betreffen, beeinflussen, beeinträchtigen	disprove	widerlegen
antimony	Antimon	do away with	beseitigen, abschaffen, aufheben
appreciable	merklich, nennenswert	be due to	wegen, infolge
ardour	Eifer, Begeisterung	endeavour	sich bemühen, streben(nach), (ver)suchen
arsenic	Arsen		
article of faith	Glaubensartikel	entire	ganz, völlig, vollständig
challenge	heraus-, auffordern, bestreiten, anzweifeln, ablehnen	entity	Wesen, Ding
		extend	ausdehnen, vergrößern, erweitern
combustion	Verbrennung		
compound	Verbindung	hinder	(ver)hindern, im Wege sein
confirm	bestätigen, bekräftigen	infer	folgern, ableiten
constituent	Bestandteil		

intellects	große Geister, kluge Köpfe	propagation	Aus-, Verbreitung, Vermehrung
investigator	(Nach)Forscher, Untersuchende(r)	provide	versehen, -sorgen, beschaffen, liefern
longevity	langes Leben, Lang-lebigkeit	by (with) reference to	bezüglich
manifest	offenbar, -kundig, augenscheinlich	refute	widerlegen
mercury	Quecksilber	remedy	Mittel, Arznei
novel	neu(artig)	research	Forschung
obscurity	Unklarheit, Undeut-lichkeit	respiration	Atmung, Atmen
owe	verdanken, schulden (Geld, Dank, Ach-tung)	ridicule	verspotten, lächerlich machen
postulate	(als gegeben) voraus-setzen, postulieren; fordern, verlangen	scope	Ausmaß, Umfang; Bereich, Gebiet
prescribe	verordnen, -schreiben	state	erklären, darlegen, anführen
previous	vorherig, vorher-, vorausgehend, früher	superstition	Aberglaube
proceed	weiter-, fortfahren	verify	(auf die Richtigkeit hin) (nach)prüfen, die Richtigkeit fest-stellen oder nachwei-sen, bestätigen
promote	fördern, unterstüt-zen, werben für		

Questions

1) What, according to the iatrochemists, was the true function of the alchemist?
2) How would iatrochemists have defined 'illness'?
3) Paracelsus and his merits. Comment!
4) Why did Boyle criticize the methods and concepts of the alchemists and iatrochem-ists?
5) Contrast the two theories concerning combustion. Why could these fundamentally different conceptions exist simultaneously over such a long period of time?
6) Why is Lavoisier considered to be the 'father of scientific chemistry'?

Grammar

The present perfect (simple and continuous)

Translate the following sentences and explain why the present perfect is used:

1) Electron beam-heating **has** long **been known.**
   Prices **have gone up.**

   Simple:

   ....

2) I **haven't seen** this apparatus **yet.**

   – till now, up to the present, so far, already, ever, (not) yet, never, lately, just, now, these days/years, etc.

3) He **has lived** in this house **since** 1981.
   He **has worn** that coat **for** five years.

   ....

4) This pupil **has failed** his final school examination at least 3 times.

   ....

1) How long **have** you **been learning** English?
   She **has been working** at that institute for about 15 years.

   Continuous:

   ....

2) I **have** (just) **been reading** a most interesting article on plastics.
   I **have been telephoning** all morning.

   ....

3) She **has been paying** a lot of bills by cheque.
   My neighbours **have been going** there every summer for the last twenty years.

   ....

Exercise

Translate into English:

1) Ich bin schon zwei Jahre an diesem Institut.
2) Er hat sie diese Woche noch nicht gesehen.
3) Diese Substanz ist seit 1967 bekannt.
4) Wie lange ist er schon an dieser Schule?
5) Ich warte schon seit drei Stunden auf ihn.
6) Er hat die Krankheit seit Weihnachten.
7) Der Versuch läuft schon 16 Minuten.
8) Ich bin seit Wochen unglücklich.
9) Sally hat ihre Bewerbung an ein chemisches Institut geschickt.
10) Sie hat bis jetzt noch keine Antwort erhalten.

Chemistry in the 19th and 20th Century

Chemistry in the 19th Century

Of course, only some of the milestones in the development of chemistry in the 19th and 20th century can be mentioned here, since the scope of this survey would otherwise be exceeded. At the beginning of the 19th century Dalton enunciated his Atomic Theory; Sir Davy and Faraday discovered the galvanic forces; Gay-Lussac announced the discovery of the important 'Law of Gaseous Volumes', and Avogadro introduced his 'principle' or 'rule' that equal volumes of different gases under identical conditions of pressure and temperature contain the same number of molecules; Wöhler succeeded in the synthesis of urea, by which, for the first time, a

product of organic life could be produced from mineral substances; paraffin, phenol, and aniline were discovered in coal tar and the first aniline dyes were produced (von Reichenbach, Unverdorben); Liebig realized the importance of mineral substances for manuring.

Systems for the classification of the elements have been proposed from early times:

the alchemists for example divided metals into 2 groups, noble and base metals.

Now, Döbereiner introduced his theory of triads; Newlands presented a table which may be regarded as the immediate forerunner of the 'Periodic Table of the Elements' by Meyer and Mendeleev (Mendeléyev/Mendeléeff/Mendelejeff).

A great step forward to solve the problem of chemical bonds was made by the dualistic electric theory of von Berzelius, the theory of radicals (Liebig) and the theory of types (Gerhardt). To organic chemistry Kekulé contributed a systematic arrangement with his theory of valence and graphic formula of carbonic hydrogens and benzene.

In addition, physical chemistry helped to interpret chemical processes. The mass action law of Guldberg and Waage enabled the calculation of chemical equilibria and the understanding of thermodynamics and energy of chemical reactions. Then the spectrum analysis of Kirchhoff and Bunsen and the discovery of osmosis (this process was first thoroughly studied in 1877 by the plant physiologist Pfeffer) extended chemical methods.

New fields of work were opened up by electrochemistry (Arrhenius' electrolytical dissociation) and colloid chemistry (Graham, Ostwald).

After Perkin's successful production of a synthetic aniline dye large scale manufacture of synthetic products began in the second half of the 19th century.

In the same period, chemical factories were founded, such as Bayer (1863), Hoechst (1863), and BASF (1865), which were to be of great importance for the later development of chemical industry.

In 1877 the 'Verein zur Wahrung der Interessen der Chemischen Industrie Deutschland e. V.' (today 'Verband der Chemischen Industrie e. V.') was established.

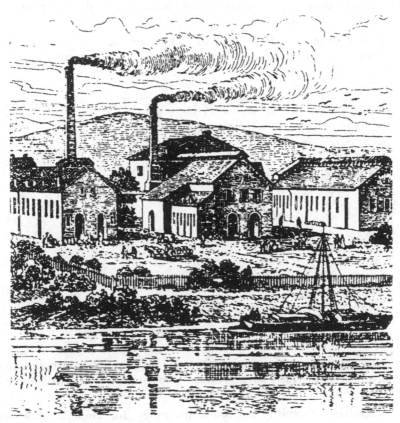

The Hoechst works in 1865

Questions

1) Take two of the discoveries made at the beginning of the 19th century and explain their use and significance in today's world.
2) Who contributed to the compilation of the 'Periodic Table'? Why is it so important to chemistry?
3) Give reasons why physical chemistry helped to explain chemical processes. Use the examples given and expand them.
4) What contributions to chemistry has electro-chemistry made?
5) Explain or give synonyms for the following words: milestone; scope; enunciate; molecule; fore-runner; thermo-dynamics; osmosis; founding; safeguard.

Chemistry in the 20th Century

Probably the greatest progress in chemistry has been made in this century, but this does not detract from the value of discoveries made in earlier times.

Considering how much has been achieved in the last eight decades it is obvious that we cannot look at everything in great detail. Another reason for not doing this is the fact that the latest discoveries are more or less known and are discussed from time to time in chemistry lessons anyway.

Important achievements of the recent past and present have very often been joint accomplishments of universities and chemical factories. Among many others we have witnessed:

> colour photography (Lumière) 1904
> ammonia synthesis (Haber, Bosch) 1908
> bakelite (Baekeland) 1909
> atomic model (Bohr) 1913
> hydrogenation of coal (Bergius) 1921
> penicillin synthesis (Fleming, Chain) 1928, 1938
> sulphonamides (Domagk) 1932
> artificial radioactivity (Joliot-Curie) 1934
> nuclear fission (Hahn) 1938
> and of course
> – synthetic rubber (buna),
> – artificial silk (rayon),
> – cellulose, and
> – a great number of plastics.

Further scientific findings made possible by the investigation into radioactivity by Marie Curie and Otto Hahn have provided chemistry with new insights into the structures, properties, and reactions of substances.

On the other hand, chemistry has provided biology, for example, with essential information on the constitution and qualities of physiologically and biologically important substances such as proteins, vitamins, and hormones.

It seems, therefore, that co-operation between the sciences, instead of each one working in isolation, will be the aim in years to come.

After this short and, of necessity, incomplete survey of the history of chemistry the question arises as to its future. It is quite clear that pressing problems will have to be solved.

Now as before, chemistry must satisfy the urgent, and sometimes less urgent, needs of mankind:

– providing food for a growing world population,
– securing a certain standard of living,
– retaining jobs and creating new ones,
– curbing pollution, etc.

Photograph of the bench where Otto Hahn did his first work on atoms (atomic fission). Deutsches Museum, München.

It is a matter of course that greater care must be taken of the important relationship between chemical industry and environment. Environmental awareness must be strengthened and chemical industry will have to spend more money on environmental protection than hitherto.

But it is also plain that neither environmental protection nor chemistry can be abandoned.

Moreover chemistry is challenged

– to find new medicines for cancer,
– to develop the 'photosynthesis in the test tube' or
– to produce materials with even better properties.

These are only a few of the tasks it will have to deal with. Some of these must be tackled first, others later. The decision as to which ones have greatest priority, does not rest with chemists alone.

One thing is certain: developments must not simply be accepted but there must be some restriction on those going in the wrong direction. Then, if need be, alternatives must be found.

As you can see, chemists and chemistry have more than one task and a great deal of responsibility for the future.

Vocabulary

accomplish-ment	Leistung, Ausführung, Realisierung	joint	gemeinsam, gemeinschaftlich
achievement	Leistung, Errungenschaft, Ausführung	large-scale (production)	in großem Maßstab, Massen-...
as to	was . . . betrifft	law of mass action	Massenwirkungsgesetz
benzene	Benzol	manuring	Düngung
bond	(Ver)Bindung	of necessity	notgedrungen, notwendigerweise
cancer	Krebs		
carbonic	Kohlen-. . .	need	Bedürfnis
classification	Einteilung, Anordnung	pressing	dringend, drückend
coal tar	Steinkohlenteer	priority	Vorrang, Dringlichkeit(sstufe)
contribute	beitragen, beisteuern		
curb	zügeln, im Zaum halten	property	Eigenschaft
		resin	Harz
enable	ermöglichen, befähigen	retain	zurück(be)halten, bewahren, (bei)behalten
enunciate	aufstellen, verkünden, ausdrücken		
		reverse	umgekehrt
environment	Umwelt	safeguard	schützen, sichern
environmental awareness	Umweltbewußtsein	safeguard interests	Interessen wahrnehmen
environmental protection	Umweltschutz	secure	sichern, garantieren, schützen
equilibrium	Gleichgewicht	silk	Seide
exceed	überschreiten, -steigen, hinausgehen über	strenghten	(ver)stärken
		survey	Überblick, -sicht; Untersuchung
finding (s)	Entdeckung, Befund	tackle	in Angriff nehmen, anpacken
forerunner	Vorläufer		
formula	Formel	task	Aufgabe
hitherto	bisher, bis jetzt	test-tube	Reagenzglas
hydrogenation	Hydrierung	triad	Triade, dreiwertiges Element
immediate	unmittelbar, direkt		
indispensable	unentbehrlich, unerläßlich	urea	Harnstoff
		urgent	dringend
insight	Einblick, Einsicht, Verständnis	valence, valency	Wertigkeit
interaction	Wechselwirkung, gegenseitige Beeinflussung		

Questions

1) Give examples to support the statement 'Probably the greatest progress in chemistry has been made in this century'.
   How do they affect our everyday life?
2) In which fields can the sciences work together. In what way is this an improvement?
3) What are the conflicts between chemical industry and the environment? How can they be resolved?
4) What tasks would you like to see accomplished by chemistry in the future? Give examples to support your answer.
5) Explain or find synonyms for the following words: achievement; radioactivity; hormone; in isolation; curb; photosynthesis; priority.

Grammar

The past tense (simple and continuous)

Study the following sentences and point out why the past tense is used:

Simple:

1) They **received** the substances three days ago.
   Yesterday, the student **experimented** for two hours. ....

2) She **went** to the stockroom, **asked** for the substances, and **took** them to the lab. ....

3) While she **was** abroad, he **wrote** to her every day. ....

Continuous:

1) What **were** you **doing** just now?
   I **was** just **testing** these plugs. ....

2) Last week they **were staying** with the Smiths. ....

3) While I **was analysing** the material, the others **were writing** their reports. ....

Exercises

A Translate into English:

1) Gestern habe ich diesen Versuch gemacht.
2) Als George heimkam, war sein Bruder gerade dabei, den Fernseher zu reparieren.
3) Im Jahre 1927 wurde die Photozelle erfunden.
4) Er hat diese Nachricht vor drei Stunden gehört.
5) Während sie sich die Hände wusch, klingelte das Telephon.

B Insert the past tense or the present perfect:

1) He (to study) chemistry for the last two years.
2) They (to be) friends for some ten years.
3) The need for more radio stations (to force) research and development into higher frequencies.
4) Before 1936 most engineers (to regard) the radio valve as a means of obtaining amplification of near-sinusoidal waveform.
5) Thus far (Up to now) the flaming arcs and the magnetic lamps (to use) to a limited extent only.
6) This device (to be) on the market for three years.
7) The sun was just setting when we (to reach) home.
8) We (to listen) to the wireless last night.
9) The students (to experiment) on these electrical devices all day long.
10) The houses (to be) built here last year.

C Translate into German:

1) These students have never visited a nuclear power station.
2) Units of 200 MW were manufactured early in 1958.
3) We were playing tennis in the garden when he phoned.
4) He has been living in Stuttgart since 1981.
5) She has had this job for four years.

2 Chemistry –
In Theory and Practice

The Matter We Breathe
Properties of Water and its Hardness
The Metals
Acids and Bases
The Synthesis of Sulphuric(VI) Acid and of Ammonia
The Chemical Laboratory
Neutralization Titration
Thin Layer Chromatography
Photometry in the Visible Range of the Spectrum
Conductivity Meter
An Electrochemical Cell
Two Frequently Applied Techniques of Separation
o-Anisaldehyde
Ethanol

The Matter We Breathe

Composition of the Air

From the time of the Greeks, for some 2000 years, air was considered to be a single substance. It was not until the 17th century that this supposition was questioned. In fact, dry air is a mixture of predominantly two gases, **oxygen** and **nitrogen.** It also contains small amounts of **carbon dioxide, argon** and other **noble gases**.

Gas	Composition by volume	Composition by mass
	%	%
Nitrogen	78.09	75.51
Oxygen	20.95	23.15
Carbon dioxide	0.03	0.04
Argon	0.93 ⎫	1.30
Other noble gases	> 0.003 ⎭	

In addition to these gases, air contains a variable amount of **water vapour,** on average 2%. Wind and industry pollute the air with **dust** and chemicals such as **sulphur dioxide, nitrogen oxides, ammonia** and many others. Man pumps into the air **smoke** from fires, **carbon monoxide** and **lead** from his car, and a variety of industrial pollutants.

The composition of the air is found to vary from place to place, confirming the fact **that air is a mixture.** Gases present in the air exhibit their own characteristic properties and can be separated by fractional distillation.

Fractional Distillation of Liquid Air

Liquid air is primarily a mixture of nitrogen and oxygen with small amounts of the noble gases. The boiling point of nitrogen is −196°C and that of oxygen is −183°C. For this reason, when liquid air is allowed to warm up, the nitrogen boils first and the remaining liquid becomes richer in oxygen. By refractionating the oxygen fraction several times, the oxygen produced is very pure. Most oxygen does not need to be so pure and is simply stored by pumping it into steel cylinders under high pressure after only one fractionation.

To prevent blockage caused by solid materials at these low temperatures, air is first freed from carbon dioxide, water vapour and dust impurities. It is then compressed to about 200 atmospheres, cooled and allowed to expand through a fine jet. This sudden expansion causes further cooling and the gas eventually liquefies. (You can observe

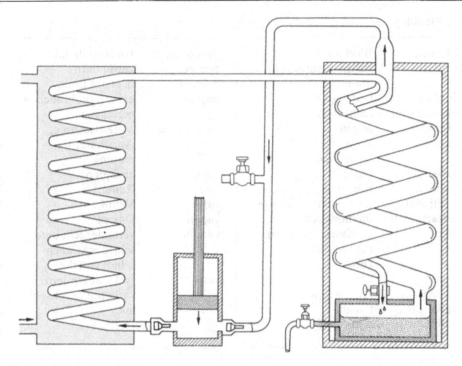

Manufacture of liquid air

this cooling effect by measuring the fall in temperature when a thermometer is placed near the opened valve of an inflated car tyre.) The liquid is tapped off through a valve, while gas which has escaped liquefaction returns to the compressor.

Vocabulary

blockage	Blockade	predominant	vorherrschend
compress	zusammendrücken,	prevent	verhindern
	-pressen	primarily	in erster Linie
confirm	bestätigen	require	erfordern; brauchen;
eventual(ly)	schließlich		verlangen
exhibit	zeigen	supposition	Voraussetzung,
expand	ausbreiten;		Annahme; Ver-
	ausdehnen		mutung
expansion	(Aus)Dehnung	tap off	abzapfen
fraction	Bruchteil	tyre	Reifen
fractional	fraktioniert, teilweise	valve	Ventil
refractionate	fraktionieren	vapour	Dampf
inflate	aufblasen, -pumpen,	variable	veränderlich, unter-
	mit Luft füllen		schiedlich, wech-
jet	Düse		selnd; variabel
lead	Blei	vary	(ver-, ab)ändern,
liquefy	verflüssigen		variieren, wechseln
noble gas	Edelgas		mit
pollutant	Schadstoff		

Questions

1) What is air composed of?
2) What does 'air pollution' mean?
3) How can air be broken down into its constituent parts?

Paraphrase:

a) supposition   b) predominant   c) primarily
d) liquefy   e) inflate   f) tap off

Grammar

The past perfect (simple and continuous)

Translate the following sentences and explain when the past perfect is used:

Simple:

1) We **thanked** him for what he **had
   done.**
   She **had** just **reached** home when it
   began to rain.
   Before they **had gone** very far, they
   found that they **had lost** their way.
   By that time it **had stopped** raining.

....

2) By November 2nd last year he **had
   lived** in this house **for** 10 years.
   I **had known** her for years before we
   were introduced to each other.   ....

Continuous:

1) How long **had** you **been learning**
   English then?
   He **had been waiting** for twenty
   minutes before she arrived.
   My brother **had been living** in that
   house ever since he came here.   ....

2) I **had** just **been reading** a most interesting article.
   He **had been having** headaches con-
   stantly for the last year.   ....

Exercises

A Supply the proper tense (present, past, present perfect, past perfect) of the verb
in parentheses:

1) He had been reading his mail when we (to arrive).
2) If it doesn't rain, we (to eat) our lunch on the porch.
3) Paula (to lie) on the floor for several hours last night listening to her records.
4) George (to work) in the same firm for ten years.
5) At the present time there (to be) three films to be shown.
6) Mr Dalf (to teach) in their school since February.
7) The twins (to swim) across the lake last summer.
8) If you hurry to the station, you (to see) her before she departs.
9) Elizabeth (to listen) to you for an hour and now she is tired.
10) Last year their films (to win) five awards out of fifteen.
11) For the last three years her son (to study) in Paris.
12) Tom (to drink) his coffee and left.
13) Yesterday it (to rain) after it (to be) dry for several months.
14) He (to come) at five o'clock as we (to arrange).
15) He (to admit) that he never (to believe) in her ability.
16) The fire (to reach) the next building before the firemen (to come).
17) He (to take) the position after they (to offer) it to him several times.
18) In the nineteenth century coal (to replace) water, which up to that time (to be)
    the chief source of power.
19) Low water, which previously (to interrupt) the work of mills on a stream, (not to
    stop) mills operated by coal.
20) This morning I (to remember) that I (to lend) Dick five dollars.

B   Translate the following text and discuss its implications:

Der Drang nach Neuem begleitet die Menschheit durch die Geschichte. Doch erst die Klärung und Anwendung naturwissenschaftlicher Zusammenhänge und Gesetze machten systematische Forschung möglich.

Die Chemie hat den Fortschritt besonders nachhaltig beeinflußt. Auf ihrer Grundlage entstand eine Industrie, ohne deren Erzeugnisse menschliches Leben heute und in Zukunft kaum vorstellbar wäre:

> Sicherung der Ernährung; Sicherung der Gesundheit; bequemes und schöneres Wohnen; pflegeleichte und preiswerte Kleidung; moderne Kommunikation; . . .

Freilich wachsen mit diesem Fortschritt auch die Probleme. Gerade die letzten zwanzig Jahre brachten neue Erkenntnisse über bisher unbekannte Gefahren bei Herstellung und Anwendung chemischer Produkte. Es gibt Abfall in verschiedener Form, mit dem die Chemie fertig werden muß.

Solche Erkenntnisse fördern, auch wenn sie manchmal mehr als nötig Unruhe schaffen, den Lernprozeß. Und deshalb ist Chemie nach mehr als hundert Jahren eben auch: mehr Erfahrung im Umgang mit chemischen Stoffen, mehr Wissen über Ursachen und Wirkungen und damit auch bessere Technik zur Entlastung der Umwelt und mehr Sicherheit der Prozeßabläufe.

Die chemische Industrie muß immer mehr ökologische Notwendigkeiten und wirtschaftliche Grenzen beachten. Das Risiko chemischer Produktion wird größer. Deshalb wird Weiterentwicklung immer mehr zum Wagnis. Daraus folgt die Aufgabe, Risiko und Wagnis in ein vertretbares Verhältnis zu Aufwand und Nutzen zu bringen, denn Entwicklungsstillstand brächte uns in Widerspruch zur geschichtlichen und biologischen Erfahrung.

Properties of Water and Its Hardness

In almost all its physical properties water is unique. It is a liquid; yet compounds similar to it and of higher relative molecular mass are gases, e. g. hydrogen sulphide. This and many other of water's special properties are attributed to the strong **hydrogen bonding** between hydrogen and oxygen atoms of adjacent water molecules.

Water has its maximum density at 4 °C. Thus ice is less dense than water and floats, whereas almost all other substances are denser in the solid state than in the liquid. Also cold water (less than 4 °C) floats on top of warmer water. Freezing therefore takes place from the surface downwards, enabling underwater life to be maintained.

At one atmosphere pressure the boiling point and freezing point of pure water are 100 °C and 0 °C, respectively. These are used as two **fixed points** in the calibration of thermometers.

Because of its **polar** nature, water is an excellent solvent for the vast majority of ionic compounds.

Causes of Hardness

(See also page 162, Chemical Formulae)

When soap forms an insoluble scum and does not easily lather with water, such water is said to be 'hard'. This **hardness** is caused by Mg^{2+} and Ca^{2+} ions dissolved in water. Note that only **soluble** compounds of magnesium and calcium can provide these ions: the usual sources are the sulphates(VI), chlorides and hydrogencarbonates.

Whereas the sulphates(VI) and chlorides occur in the Earth's crust, hydrogencarbonate is produced when rain-water containing dissolved carbon dioxide passes through calcium(II) carbonate rocks, e. g. chalk and limestone. The dilute solution of carbonic acid (carbon dioxide dissolved in water) reacts with the calcium(II) carbonate to form **soluble** calcium(II) hydrogencarbonate:

$$Ca^{2+}CO_{3(s)}^{2} + \underset{\text{carbonic acid}}{(H_2O_{(l)} + CO_{2(g)}} \rightarrow Ca^{2+}(HCO_3^-)_{2(aq)}$$

Calcium(II) carbonate (chalk etc.) does not **directly** cause hardness since it will not physically dissolve in water.

Vocabulary

adjacent	angrenzend, -liegend; benachbart	insoluble	unlöslich
attribute	zuschreiben; zurück- führen	lather	schäumen
		limestone	Kalkstein
calibration	Kalibrierung, Eichung	maintain	(aufrecht)erhalten; bewahren
chalk	Kreide	occur	vorkommen
compound	Verbindung	respectively	beziehungsweise
density	Dichte	scum	Schaum
dense	dicht	solvent	Lösungsmittel
dilute	verdünnt	unique	einzig(artig); außergewöhnlich
dissolve	(auf)lösen	whereas	wohingegen
enable	ermöglichen		

Questions

1) What is the construction of a water molecule?
2) What do we understand by 'hardness of water'?
3) What causes hardness?
4) Why is water so unique and essential?

Explain:

a) unique   b) attribute   c) adjacent   d) Earth's crust

Grammar

Forms of the future

Have a look at the various forms that can be used to express future activities or states and form sentences:

1) **shall/will:**   a) ....

                                            b) ....

                                            ....

2) **to be going to:**   a) ....

                                            ....

3) **the present continuous:**   a) My father is arriving at four o'clock
   (. . . is arriving . . .)   in the afternoon.

                                            b) ....

                                            ....

4) **shall/will + continuous form:**   a) ....
   (. . . will be running . . .)

                                            ....

                                            b) ....

                                            ....

5) **the present simple:**   a) ....
   (. . . leaves . . .)

                                            ....

                                            b) ....

                                            ....

6) **the future perfect:**   a) ....
   (. . . shall have read . . .)

                                            ....

                                            b) ....

                                            ....

Now explain when the various forms are used:

1) ....

2) ....

3) ....

4) ....

5) ....

6) ....

The Metals

What are metals?

Aluminium, iron, tin, copper and lead are very much part of the world we live in. Most people recognize these substances as metals by their hardness and in their shiny lustrous surface. They can be hammered into shape **(malleability)** and drawn into wires **(ductility).** These are only a few of the physical properties which the chemist uses to classify elements as metals. In order to explain these physical properties we must look again at the structure of metals. A useful model of a metal is one in which the atoms exist as spherical, positive ions (cations) arranged in a regular three-dimensional network or 'crystal lattice'. The electrons present in the metal are distributed in such a way that the groups of positive metal ions in the crystal lattice are surrounded by a 'sea' of mobile electrons. These mobile electrons constitute the **metallic bond**.

Metallic Properties and the Metallic Bond

Conductivity

High **electrical conductivity** in metals results from the ease of the electron movement from one place in the metal crystal to another.

High **thermal conductivity** in metals is largely due to the movement of their mobile electrons.

Metallic lustre

Metals are crystalline and their flat crystal surfaces are able to reflect light. This may also be explained, in part, by the surface mobile electrons absorbing and re-emitting light energy.

Malleability and ductility

In contrast to ionic and covalent crystals, which are **brittle,** metals have high malleability and ductility because it is relatively easy for metal atoms to be moved about within the lattice without destroying the bond.

Vocabulary

bond	Bindung	lattice	Gitter
brittle	spröde, zerbrechlich; brüchig	lustre	Glanz
		lustrous	glänzend, strahlend
constitute	ausmachen, bilden	malleability	Verformbarkeit
conductivity	Leitfähigkeit	mobile	beweglich
covalency	Kovalenz, Atombindung	network	Netz
		recognition	Erkennen
ductility	Dehn-, Streckbarkeit	re-emit	wiederaussenden
distribute	verteilen	spherical	kugelförmig
due to	wegen	wire	Draht
ease	Leichtigkeit		

Questions

1) What are the characteristics of metals?
2) How are they composed?
3) What are they used for?

Paraphrase

a) spherical ions   b) lattice   c) absorb and re-emit

Grammar

Modal Auxiliaries

Modal or defective auxiliaries express such concepts as
permission, possibility, ability, (strong) obligation, probability, certainty, necessity, willingness, desirability, volition, (polite) request, (polite) suggestion, promise, duty, consequence, etc.

Attach one or more of the following common modals to the concepts given above:
must, may, can, will, shall,
would, should, ought to, might, need, used to.

What are the differences between modals and full verbs?

– modals have no -s in the third form singular, present tense,

– ....

– ....

– ....

– ....

Insert the corresponding, principal substitute(s) and the negative form(s) of each modal:

can:	to be able to	can't, cannot, not to be able to
may:	....	....
must:	....	....
shall:	....	....
will:	....	....

Exercises

A   Form sentences with the different meanings these modals can have, put them into other tenses and form the negative form.

can

to be able to (ability):   Paula can carry out this experiment on her own.
                                         Paula had been able to carry out this experiment on her own.
                                         Paula cannot carry out this experiment on her own.

to be allowed to (permission)   ....

it is possible to (possibility)   ....

may
to be allowed to (permission)   ....

maybe, perhaps (possibility)   ....

must
to have to (necessity, obligation)   ....

to be certain (certainty)   ....

shall
to be to (instruction, arrangement)   ....

will
to want to (volition)   ....

B   Put into the tenses suggested:

1) She mustn't open this carboy (Past Tense).
2) Our environment must be protected (Future Tense).
3) Each student may take 7 g of this substance (Present Perfect).
4) Can you solve this equation? (Future Tense).
5) Students cannot smoke in the lab (Past Perfect).

C   Translate into English:

1) Sie hatte diese Analyse nicht selbständig durchführen können.
2) Im Labor werden immer Schutzbrillen getragen werden müssen.
3) Die Schüler durften dieses Experiment nicht alleine machen.
4) Er sollte diese Arbeit bis morgen tun.
5) Einige Substanzen dürfen nicht dem Licht ausgesetzt werden.

Acids and Bases

Svante Arrhenius (1859–1927), a Swedish scientist, received the Nobel Prize in Chemistry in 1903 for his work on the dissociation theory of acids, bases and salts. Arrhenius postulated the existence of ions in aqueous solutions. As early as 1884 he suggested that **substances which yield hydrogen ions in aqueous solution are acids.** Hydrogen chloride gas yields hydrogen ions in aqueous solution as follows:

$$HCl_{(g)} \rightarrow H^+_{(aq)} + Cl^-_{(aq)}$$

The solution is an acid; it is in fact hydrochloric acid. Similarly, nitric(V) acid yields hydrogen ions in solution:

$$HNO_{3(l)} \rightarrow H^+_{(aq)} + NO^-_{3(aq)}$$

Substances giving hydroxide ions, OH^-, **in aqueous solution are bases** according to Arrhenius' theory. For example, sodium(I) hydroxide and potassium(I) hydroxide yield hydroxide ions as follows:

$$NaOH_{(s)} \rightarrow Na^+_{(aq)} + OH^-_{(aq)}$$
$$KOH_{(s)} \rightarrow K^+_{(aq)} + OH^-_{(aq)}$$

A **neutralization reaction** between any acid (a source of $H^+_{(aq)}$) and any alkali (a source of $OH^-_{(aq)}$) is a combination of hydrogen ions and hydroxide ions to form water molecules, and can be written:

$$H^+_{(aq)} + OH^-_{(aq)} \rightarrow H_2O_{(l)}$$

One of the problems faced by Brønsted in Denmark and Lewis in England (1923) was the nature of the hydrogen ion $H^+_{(aq)}$. This problem can be illustrated by looking closely at the acidity of hydrogen chloride. If dry hydrogen chloride gas is dissolved in dry methylbenzene (toluene), the solution shows no acidic properties and is non-conducting. No ions are present in this solution.

When hydrogen chloride is dissolved in water, the solution gives an acid reaction with carbonates and allows an electric current to pass through it with decomposition taking place at the electrodes.

If the solution of hydrogen chloride in methylbenzene is shaken with water, the methylbenzene and water form two layers. The lower aqueous layer displays all the properties of the hydrogen-chloride/water solution, showing that some of the hydrogen chloride has passed into the aqueous layer. Thus the water must play some part in the acidity of hydrochloric acid. We believe that the hydrogen ion is hydrated by a water molecule to form an *oxonium* ion $H_3O^+_{(aq)}$

$$H^+_{(aq)} + H_2O_{(l)} \rightarrow H_3O^+_{(aq)}$$

and that hydrogen chloride dissolves in water as follows:

$$HCl_{(g)} + H_2O_{(l)} \rightarrow H_3O^+_{(aq)} + Cl^-_{(aq)}$$

This is in line with the Brønsted-Lewis definition of an acid as **any substance consisting of molecules or ions that donate protons.** In the above reaction the hydrogen chloride donates a proton (H^+) to a water molecule and is therefore an acid. Any compound releasing hydroxide ions must be a base, because the hydroxide ion is capable of accepting a proton to form water:

$$OH^-_{(aq)} + H_3O^+_{(aq)} \rightarrow 2H_2O_{(l)}$$

Similarly, ammonia is a base capable of accepting protons:

$$NH_{3(g)} + H_2O_{(l)} \rightarrow NH^+_{4(aq)} + OH^-_{(aq)}$$

The base (NH_3) accepts a proton from the water.

Although we appreciate that an acid donates a proton to a water molecule it is often convenient to write the hydrated proton simply as $H^+_{(aq)}$.

Vocabulary

acidity	Säuregehalt, Säure, Schärfe	display	zeigen, offenbaren
		donate	abgeben
acidic	sauer	in line with	übereinstimmen
appreciate	(richtig) beurteilen, erkennen; schätzen	layer	Schicht
		non-con-ducting	nicht-leitend
aqueous	wäßrig		
capable	fähig, imstande	postulate	postulieren, (als gegeben) voraus-setzen
convenient	bequem, geeignet, (zweck)dienlich, praktisch		
		take place	stattfinden
decomposi-tion	Zersetzung, Aufspal-tung	yield	liefern

Questions

1) What are acids? What are bases? Give examples.
2) What do we understand by 'dissociation'?
3) What is the oxonium ion and how is it formed?

Paraphrase:

a) postulate   b) yield   c) donate
d) appreciate   e) convenient

| Grammar |

Describing
dimensions, objects, shape, size, use

Height	– high
width	– wide
breadth	– broad
length	– long
depth	– deep

Height, width, etc. are **linear** dimensions!

Now, look around and use these patterns to ask and answer questions about objects in your classroom, e.g. table, books, lamps, etc.

What is the	height breadth width	of .... ?
	....	

The	height breadth	of .... is ....
	....	

.... is .... in	height. width.
	....

.... has a	height length	of ....
	....	

How	long wide	is .... ?
	....	

.... is	long. wide.
	....

When asking and answering questions about various objects, use the following tables:

How	heavy large big	is it?

What	size shape colour	is it?

It's	very rather fairly not very quite extremely	large. small. big. light. ....

It's	dark light	blue. green. brown. ....

What does it	contain? consist of?
What's it	made of? used for?

It contains ....

It's used for .... ing ....

It's made of ....

It consists of ....

It's		approximately about roughly	20 kg.
It	weighs measures	almost nearly	6 qm. 30 cm.
		just over just under	etc.

Exercise

Now, test yourself:

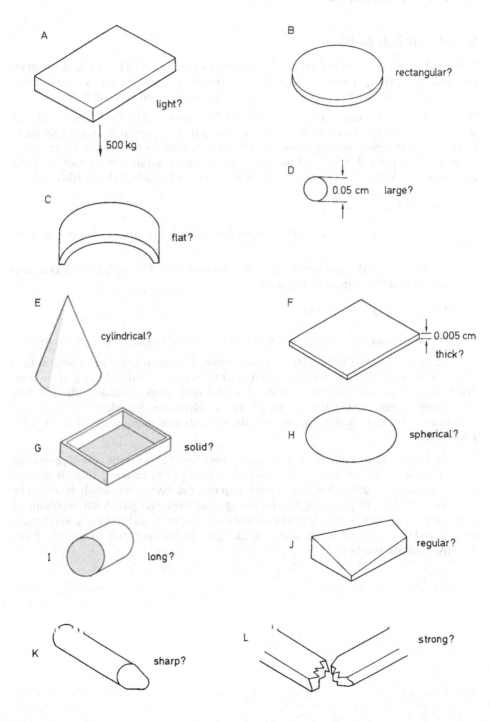

A   light?

500 kg

B   rectangular?

C   flat?

D   0.05 cm   large?

E   cylindrical?

F   0.005 cm   thick?

G   solid?

H   spherical?

I   long?

J   regular?

K   sharp?

L   strong?

The Synthesis of Sulphuric(VI) Acid and of Ammonia

Sulphuric(VI) Acid

The most important method for the manufacture of sulphuric(VI) acid is the **contact process.** The principal raw material for this process is native sulphur or, to a lesser extent, some sulphur-containing compound such as pyrites (an iron sulphide).

Sulphur is burned in excess dry air to form sulphur dioxide. The gas mixture (sulphur dioxide and air) is filtered to free it from dust and then passed through a series of converters containing trays of vanadium(V) oxide (vanadium pentoxide) which acts as a catalyst. Sulphur dioxide and oxygen from the air combine on contact with the catalyst in an exothermic reaction to produce sulphur(VI) oxide (sulphur trioxide):

$$2SO_{2(g)} + O_{2(g)} \rightarrow 2SO_{3(g)}$$

Cooling air is used to maintain the temperature in the converters at approximately 450°C.

The sulphur(VI) oxide (sulphur trioxide) is absorbed in 98–99% sulphuric(VI) acid to form **oleum** (fuming sulphuric(VI) acid):

$$H_2SO_{4(l)} + SO_{3(g)} \rightarrow H_2S_2O_{7(l)}$$

Suitable dilution of the oleum gives sulphuric(VI) acid of any desired concentration.

In the contact process the catalyst is **surface active**. This means that gas reactions take place on the surface of the catalyst, which must therefore provide a large surface area. Dust will reduce the effective surface area and may even react with the catalyst, 'poisoning' it and further limiting its efficiency. However, vanadium(V) oxide is a reasonably efficient catalyst for the oxidation of sulphur dioxide and is not readily poisoned.

In a sulphuric(VI) acid plant one of the most important factors is the cost of producing pure, **dust-free** sulphur dioxide. Although it is cheaper to produce sulphur dioxide from sulphur ores rather than from native sulphur, the saving may easily be offset by the extra expense of purifying and removing dust from the gas. Such problems of economics are common in industrial chemistry and the manufacturer must weigh them carefully. He may well decide that it is cheaper in the long run to use the more expensive native sulphur.

Vocabulary

catalyst	Katalysator	offset (v.)	ausgleichen, aufwiegen, wettmachen
contact process	Kontaktverfahren		
		oleum	Oleum, rauchende Schwefelsäure
excess	Überschuß, Übermaß		
		principal	hauptsächlich, Haupt. . .
expense	Ausgabe, Aufwand		
extent	Umfang, (Aus)Maß, Grad	pyrite	Pyrit, Schwefel-, Eisenkies
in the long run	auf die Dauer	saving	Einsparung
native	(ein)heimisch	tray	Schale

Questions

1) Describe the production of H_2SO_4 in the contact process.
2) What is H_2SO_4 used for?
3) What do we understand by 'oleum'?

Paraphrase:

a) manufacture   b) excess   c) tray
d) catalyst   e) offset   f) fertilizers

Ammonia

The synthesis of ammonia from nitrogen and hydrogen is carried out in the **Haber-Bosch-process:**

$$N_{2(g)} + 3\,H_{2(g)} \rightleftharpoons 2\,NH_{3(g)}$$

This is an exothermic reaction and under normal conditions of temperature and pressure it is extremely slow. To increase the rate of the reaction, finely divided iron (or tungsten) is used as a catalyst. (Note that the introduction of a catalyst also increases the rate of the **reverse** reaction; however, it enables the equilibrium position to be reached more rapidly.) The optimum conditions from the point of view of cost and efficiency are a moderately high temperature of 500°C and a pressure of 150–300 bars. Under these conditions the equilibrium mixture contains approximately 15% of ammonia. The ammonia is removed from the unchanged nitrogen and hydrogen by liquefaction, and unchanged gases are recycled over the catalyst.

Often almost all of the hydrogen for the ammonia synthesis is obtained from natural gas (methane) by reaction with steam, using a nickel catalyst:

$$CH_{4(g)} + H_2O_{(g)} \longrightarrow CO_{(g)} + 3\,H_{2(g)}$$

Some hydrogen is still produced from water gas (a mixture of hydrogen and carbon monoxide) by mixing it with steam and passing it over an iron(III) oxide catalyst at 400 °C:

$$\underbrace{H_{2(g)} + CO_{(g)}}_{\text{water gas}} + H_2O_{(g)} \longrightarrow 2\,H_{2(g)} + CO_{2(g)}$$

water gas

The carbon dioxide is removed by washing the gas mixture with water under pressure.

Nitrogen for the Haber-Bosch process is obtained from the air.

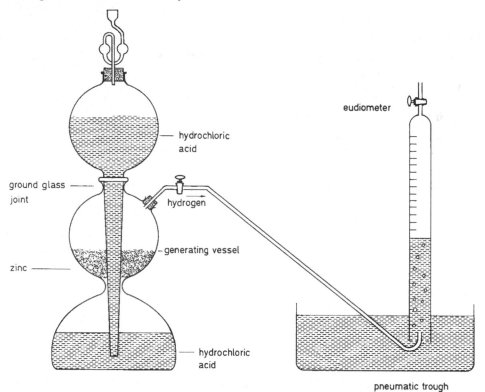

eudiometer

hydrochloric acid

ground glass joint

hydrogen

generating vessel

zinc

hydrochloric acid

pneumatic trough

Vocabulary

approximate(ly)	annähernd	methane	Methan
		moderate	mäßig; angemessen
equilibrium	Gleichgewicht, Balance	reverse	umgekehrt, entgegengesetzt
liquefaction	Verflüssigung	tungsten	Wolfram

Questions

1) How is ammonia produced on a large scale?
2) On what conditions does the yield depend? Why?
3) How are hydrogen and nitrogen (necessary for this process) obtained?

Explain:

a) exothermic   b) equilibrium   c) recycle

Exercises

Describe the diagram

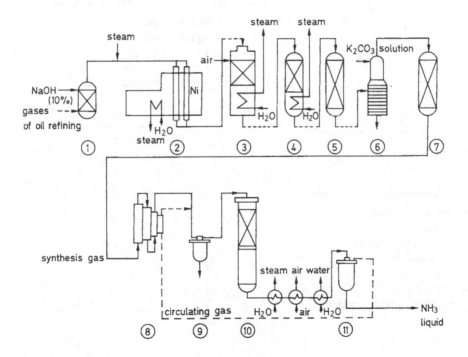

Technological flow sheet of the ammonia synthesis (simplified)

① caustic washing   ⑥ absorber
② primary reformer   ⑦ methanisator
③ secondary reformer   ⑧ compressor
④ converter I   ⑨, ⑪ ammonia separator
⑤ converter II   ⑩ ammonia reactor

Objects

Which belongs to what?

1) depth
2) hollow
3) solid
4) thickness
5) diameter (d)
6) radius (r)
7) circumference (c)

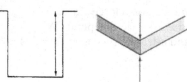

A.

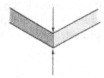

B.

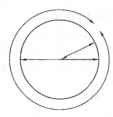

C.

D.

E.

A.

B.

C.

D.

E.

What are the names of these shapes?

1) Elliptical plate
2) Square plate
3) Semicircular plate
4) Triangular plate
5) Rectangular plate

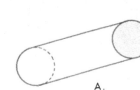

A.

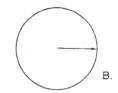

B.

Give each shape its proper name.

1) Pyramid (pyramidal)
2) Ellipse (ellipsoidal)
3) Cylinder (cylindrical)
4) Cone (conical)
5) Cube (cubic)
6) Sphere (spherical)
7) Hemisphere (hemispherical)

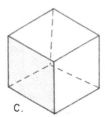

C.

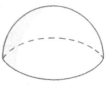

D.

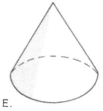

E.

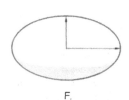

F.

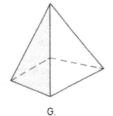

G.

Here we have a .... line.

" .... line.

" .... line.

" .... line.

" .... line.

" .... line.

(dotted, pointed, curved (2), rounded, straight, wavy, broken, zigzag, flat)

This rod is .... at one end.

This rod is .... at one end.

This plate is ....

This plate is ....

Information Given by an Equation

Consider the following balanced equation:

$$H_{2(g)} + Cl_{2(g)} \rightarrow 2HCl_{(g)}$$

This equation tells us that one mole of gaseous hydrogen reacts with one mole of gaseous chlorine to produce two moles of gaseous hydrogen chloride. In general, any balanced chemical equation contains the following information:

a) the reactants and the products;
b) the relative number of moles of each product and each reactant;
c) the physical state of the reactants and products.

However, a balanced equation does **not** tell us:

1) the conditions necessary for the reaction to take place;
2) the concentration of the reactants;
3) the reaction rate, i. e. how quickly the reaction proceeds;
4) the extent or completeness of the reaction, i. e. whether or not all the reactants will be converted into products;
5) the mechanism by which the reaction takes place;
6) the energy changes which occur during the reaction.

Consider the following balanced equation for the reaction between magnesium metal and hydrochloric acid:

$$Mg_{(s)} + 2H^+Cl^-_{(aq)} \rightarrow Mg^{2+}Cl^-_{2(aq)} + H_{2(g)}$$

This equation does not tell us that the reaction is spontaneous when the magnesium is added to the acid; that the reaction is rapid if the acid is concentrated or gives any indication of the rate or extent of the reaction. There is no suggestion in the equation as to the mechanism by which the reaction proceeds, nor any evidence that heat is produced as the magnesium dissolves.

Vocabulary

evidence	Beweis; Zeichen	proceed	vor sich gehen, von-
gaseous	gasförmig		statten gehen; fort-
indication	Anzeige;		schreiten, -fahren
	(An)Zeichen,	spontaneous	spontan
	Anhaltspunkt		

Calculations Involving Reaction Masses

The chemist in industry is constantly using chemical equations to calculate reacting quantities. He will want to know, for example, the amount of raw materials he must obtain for the manufacture of a specified amount of product. The method of calculation is the same whether the reaction is complex or simple, and is illustrated in the following example:

A student heats a mixture of iron filings and sulphur to produce iron(II) sulphide. What mass of iron will react with 10 g of sulphur, and what mass of iron(II) sulphide will be formed? Relative atomic masses of iron and sulphur are 56 and 32, respectively.

Solution. The balanced equation for the reaction is

$$Fe_{(s)} + S_{(s)} \longrightarrow Fe^{2+} + S^{2-}_{(s)}$$

| 56 | 32 | | 56 + 32 |

$$88$$

$\therefore$ 56 g iron reacts with 32 g sulphur to produce   88 g iron(II) sulphide

$\therefore \dfrac{56}{32}$ g iron reacts with 1 g sulphur to produce $\dfrac{88}{32}$ g iron(II) sulphide

$\therefore \dfrac{56}{32} \times 10$ g iron reacts with 10 g sulphur to produce $\dfrac{88}{32} \times 10$ g iron(II) sulphide

Thus the reacting mass of iron is $\dfrac{56}{32} \times 10 = 17.5$ g, and the mass of iron(II) sulphide formed is $\dfrac{88}{32} \times 10 = 27.5$ g.

Vocabulary

| iron filings | Eisenfeilspäne | | obtain | erhalten |

Questions

1) For what reasons are reaction equations formulated?
2) Say in your own words what can and cannot be inferred from a reaction equation?
3) Complete this equation:

$$Ba^{2+} + Cr_2O_7^{2-} + H_2O \longrightarrow BaCrO_4\!\downarrow + H^+$$

Explain:

a) balanced reaction   b) proceed   c) convert
d) occur   e) spontaneous   f) evidence
g) amount   h) iron filings

Exercises

Describing
angles and lines

a) This line is

   ....

b) This line is   _____

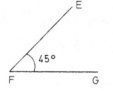

c) The line AB is .... to the line XY.

d) EF is ....of 45° to FG.

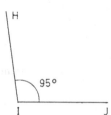

e) HIJ is a 95° angle. HIJ is an angle of 95°.
   HI is at an angle of 95° to IJ.

Note: An angle less than (<) 90° is called an **acute angle.**
An angle greater than (>) 90° is called an **obtuse angle.**
An angle greater than (>) 180° is called a **reflex angle**
or **external angle.**

Now. describe the following angles:

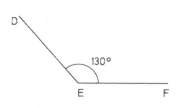

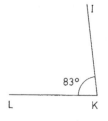

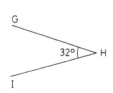

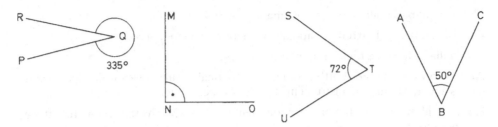

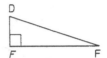

Draw the following triangles:

a) The **equilateral** triangle:

b) The **right-angled** triangle:

c) The **isosceles** triangle:

d) The **obtuse-angled** triangle:

e) The **acute-angled** triangle:

The Chemical Laboratory

To a layman, when entering a lab for the first time, the chemical laboratory is a slightly frightening and strange place. This is due to the fact that he is faced with objects and phenomena that are new and unfamiliar to him, sometimes even repulsive.

Foto: Waldner Laboreinrichtungen[13]

There might be a smell in the air that makes him recoil: it stinks.

His vision might be blurred by vapours and fumes filling the room.

Unfamiliar noises of a highly intimidating nature.

And everywhere he sees bottles containing indefinable substances which have exotic names and apparatus and instruments he has never come across before.

He is afraid of inflammable or poisonous materials, and he has the feeling that things are about to explode or react violently.

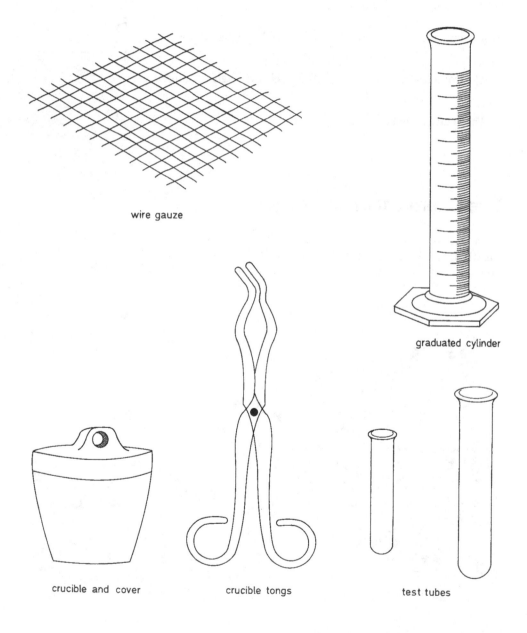

wire gauze

graduated cylinder

crucible and cover   crucible tongs   test tubes

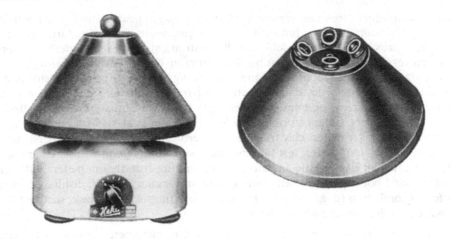

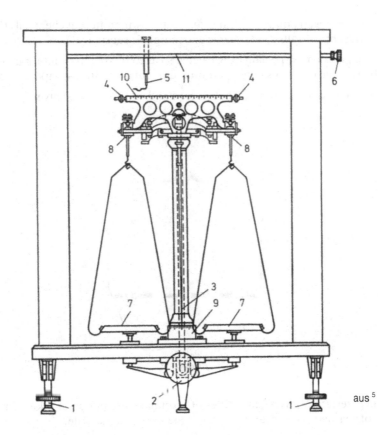

aus [5]

But close up, things aren't as unpredictable as they may appear. Usually, a lab is fitted out with laboratory benches, fume cupboards, drying ovens and muffle furnaces, sinks and cupboards containing the following lab instruments for each student: beakers, Erlenmeyer flasks, burners (microburners, Bunsen burners, Méker burners etc.), watch glasses, funnels (separatory funnel, dropping funnel), tripods, asbestos plates (boards), wire gauzes, desiccator, suction apparatus with tulip and rubber joint, suction and gas tubes, pipettes, burettes, (plastic) wash bottles, measuring flasks, graduated cylinders, (filter) crucibles (made of porcelain or glass) with lids and shoes, crucible tongs, filter paper, clay triangles, rubber wipers, test tubes, stoppers, racks, holders, clamps, spot plates, test tube brushes, winchesters (with various chemicals), pH-paper, cobalt glass, pestle and mortar, evaporating dish, thermometers, different flasks (round-bottomed fl.), fractionating column, vacuum adapter, distillation head, different condensers (e. g. Liebig c., reflux c.), cork rings, suction flask, suction filter, Woulfe's bottle, stirring rods, and so on.

There are also stands, centrifuges, spectroscopes, refractometer, water jet injector, laboratory balance, indicators and acids and bases (in hoods) at the student's disposal.

In addition to these instruments each bench has water and gas taps and a source of electricity.

In the stockroom one finds tins, cans, and bottles with different chemicals (liquids and solids, acids, bases, salts, and organic and inorganic substances).

The weighing room is specially constructed so as to be free from internal and external disturbances such as drops in temperature, draughts, vibrations, vapours, etc.

Usually, the balances (analytical balances) stand on firm stone shelves.

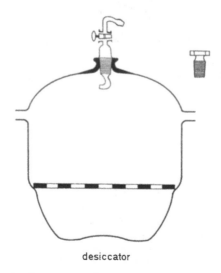

desiccator

Mostly the materials and substances to be weighed are taken to and from the balance room in a desiccator. Small quantities of substances are put on or taken off the pan by means of spatulas. The accurate weight is then read on the scale.

In addition to the rooms mentioned before, you can very often find a library, a workshop, and a room for glass blowing.

Vocabulary

beaker	Becherglas	layman	Laie
blur	verschwommen	mortar	Mörser
	machen; trüben	muffle furnace	Muffelofen
clamp	Klammer, Zwinge	pestle	Pistill
clay triangle	Tondreieck	porcelain	Porzellan
close up	näher betrachtet	rack	Gestell, Rahmen,
crucible	(Schmelz)Tiegel		Halter
crucible tongs	Tiegelzange	recoil	zurückprallen
desiccator	Exsikkator,	repulsive	abstoßend
	Trockenapparat	rubber joint	Gummiverbindung-
disposal	Verfügung; Beseitigung		(sstelle)
disturbance	Störung	slight	leicht, schwach
draught	Luftzug	spatula	Spatel
evaporating	Abdampfschale	stirring rod	Glasstab
dish		stockroom	Lager(raum)
flask	Kolben	stopper	Stöpsel, Pfropfen
fractionating	Fraktionierkolonne	suction	Saugapparat
column		apparatus	
fume cupboard	Abzug(sschrank)	suction filter	Saugfilter
funnel	Trichter (Tropf-,	suction flask	Absaugflasche
(dropping,	Scheide-)	tripod	Dreifuß
separatory)		tulip	Tulip
hood	Abzug	unpredictable	unvorhersehbar
injector	Strahlpumpe	watch glass	Uhrglas
intimidate	einschüchtern,	winchester	Vorratsflasche für
	abschrecken		Flüssigkeiten
laboratory	Laborwaage	wire gauze	Draht-, Metallgaze,
balance			-gewebe

Exercises

Explain what certain instruments, tools, etc. are used for. Separatory funnel, asbestos board, crucible tongs, etc.

Paraphrase:

a) slight
b) repulsive
c) recoil
d) blurred
e) intimidate
f) close up
g) unpredictable

Safety Precautions

In a laboratory, safety precautions should strictly be observed to avoid serious injuries to you and your laboratory neighbours. Use the following words and form whole sentences:

1) goggles/ other protective eye equipment, to wear; contact lenses/regular glasses, no substitute for.
2) Protective lab aprons/coats, gloves, necessary.
3) To know, location, fire extinguishers/protective equipment/eye-washing fountains, showers/fire blankets.
4) Not to taste, anything.
5) Dosimeters, gas masks, to detect/against vapours etc.
6) Instructions, to read, carefully.
7) Never, to carry out, unauthorized experiments.
8) To keep clean, table tops (to look like, drop of water, might be, in fact, strong acid/base solution).
9) Liquid chemicals, from the stockroom, plastic bucket, to carry, such containers, between stockroom and lab.
10) Earthing/grounding (Am.), tin-cans, to pour into another vessel, inflammable solvents.
11) To know, directions/instructions of, professional association.
12) Farsightedness/anticipation: potential dangers, to foresee; preventive measures; to have available, or, to know where things are, e. g. first aid equipment; to know how to use them.
13) Not to transfer or measure by means of a pipette, with your mouth.
14) No chemicals, on your skin; gloves.

Now carry on and state further precautions that should be taken when working in a lab.

. . . .

. . . .

Vocabulary

anticipation	Voraussicht, Vorge-fühl	preventive	vorbeugend, Vorbeu-gungs . . .
apron	(Schutz)Schürze	table top	Tischplatte
detect	feststellen; entdecken	transfer	(hinüber)bringen,
earthing	Erdung		-schaffen (von . . .
farsightedness	Umsicht		nach)
goggles	Schutzbrille	vessel	Gefäß
precaution	Vorsichtsmaßregel, Vorkehrung		

Exercises

Describing
movement and **action**

What words do you use to describe these actions?

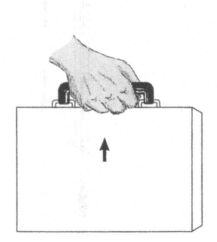
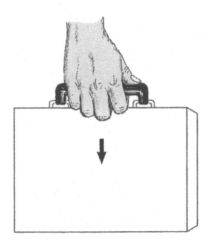

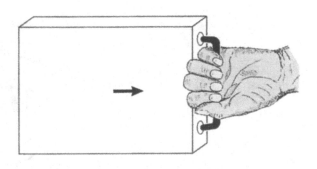

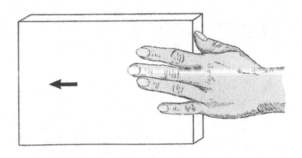

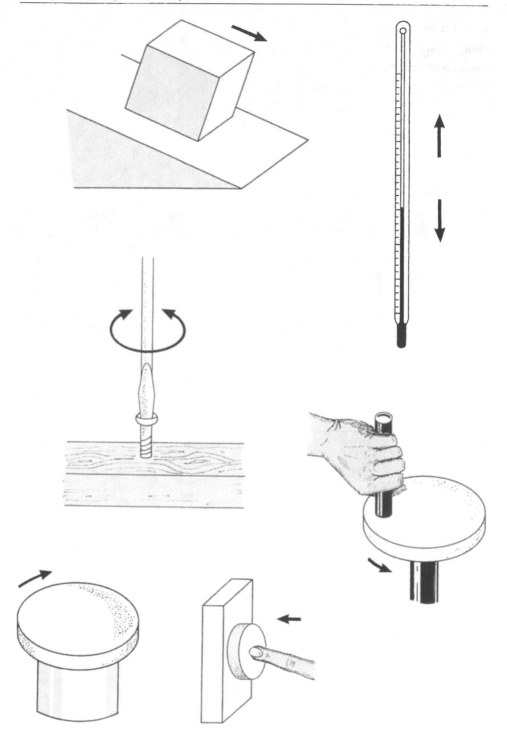

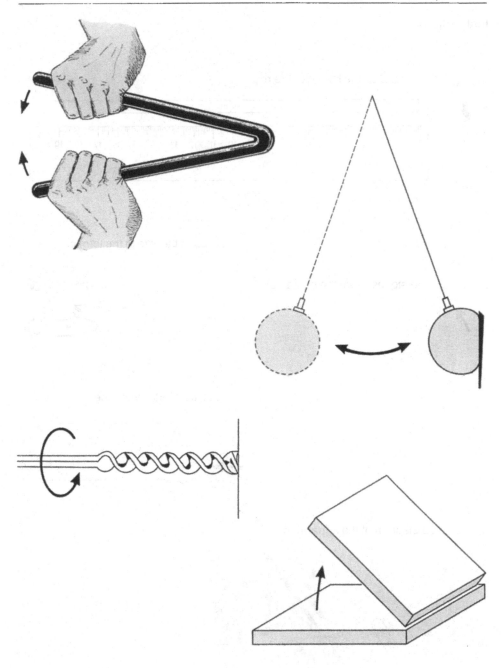

(bend, lower, swing, turn/rotate/twist/revolve, press, pull, oscillate, push, tighten, loosen, raise/lift, slide/slip, tilt).

Instructions

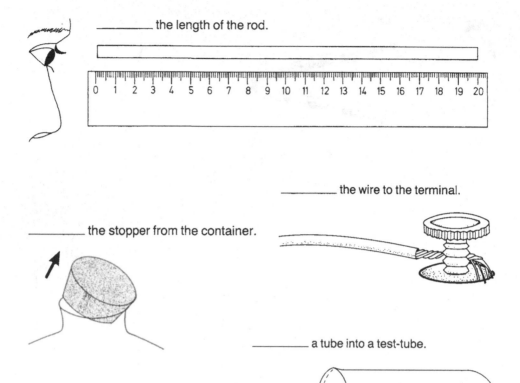

_____ the length of the rod.

_____ the wire to the terminal.

_____ the stopper from the container.

_____ a tube into a test-tube.

_____ a diagram of the apparatus.

Neutralization Titration

Volumetric Determination of the Chloride Ion

To the quarter of the analysis, diluted to 150 ml, three drops of phenolphthalein are added (0,1 g p-dioxy-diphenylphthalid in 100 ml 20% methyl alcohol).

The indicator is present in the solution in its acid form (colourless). In the range where the colour of the indicator is changing, phenolphthalein is slightly pink, and red in alkaline solution.

The interval of the colour change is in the pH-range of 8.2–10.0.

$$H^+ + Cl^- + Na^+ + OH^- \rightarrow Na^+ + Cl^- + H_2O$$

With the soda solution, which is added drop by drop, the salt acid forms neutral sodium chloride. In this way, the degree of acidity decreases and the pH-valence increases.

The solution is neutral once we have added the same quantity of soda solution, which is equivalent to the former quantity of hydrochloric acid. (This means that the end point of the titration has been reached).

As soon as the first drop of excess NaOH solution is added, the indicator changes to a slight pink.

The applied NaOH is either exactly 0.1 or has a predetermined factor. (Determination of the factor of the solution of NaOH solution).

Read off the applied ml of NaOH solution on the graduated burette.

Vocabulary

equivalent	gleichwertig	predeter-	vorher bestimmen,
decrease	abnehmen	mine	vorher festsetzen
increase	zunehmen, ansteigen	range	Bereich

Questions

1) What measuring solution is used?
2) How do you produce it?
3) What is an original titer? Give examples.
4) What does 'standardizing of a measuring solution' mean?
5) Draw the titration curve for hydrochloric acid with potassium hydroxide.

Paraphrase:

a) indicator   b) equivalent   c) predetermined

Exercises

Describing
an experiment

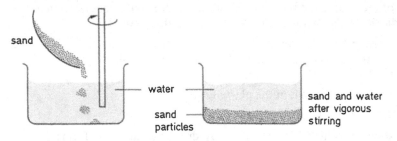

Two beakers are filled with water, and into one a quantity of sugar is stirred. A quantity of sand is placed in the other, and ....

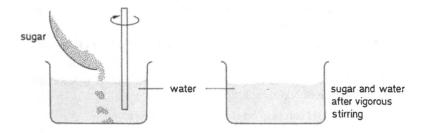

solution

Use the following words: disappear, cease, particles, grains, settle, bottom, not separate it out from, disperse; dissolve, saturated

Now describe the following experiments:

1) I take some chalk and stir it into some water in a beaker. It ....

2) Now I do the same with some salt. It ....

Gravimetric Determination of the Chloride Ion

Fundamentals

Theoretically we can use several cations for the precipitation of the chloride ion, but only the precipitation as a chloride of silver has prevailed.

The solubility product is $1.1 \cdot 10E{-}10 \, mol^2 l^{-2}$,

which, however, depends largely on the type of precipitate.

The precipitate may be flocculent, spongy, or crystalline. Moreover, the presence of ammonia or alkali salt and the acid concentration of the analysis solution have an effect on the solubility.

At first the chloride of silver forms colloidally. When a slight excess of Ag^+ ions is reached, the solution has to be heated and stirred vigorously in order to conglomerate the precipitate, which can then conveniently be filtered.

The determination is very exact and has therefore been used to determine atomic weights.

Please note that the solubility at first decreases correspondingly to the law of mass action in accordance with the excess of Ag^+ ions. But when the Ag^+ ion concentration becomes too high, the chloride of silver may dissolve again as the complex salt $AgCl \cdot AgNO_3$. Since silver halide slowly disintegrates as a result of the action of light, we have to protect the precipitate from bright daylight and UV-light. Otherwise, the precipitate will turn lilac, violet, and finally slate grey.

This change of colours is caused by the splitting of $AgCl$ into chlorine gas and colloidal silver, which is black. It is possible to protect the precipitate to a certain extent by adding methyl orange.

Vocabulary

conglomerate	zusammengeballt	prevail	vorherrschen
flocculent	flockig	slate	Schiefer
in accordance with	in Übereinstimmung mit	spongy	schwammig, porös; locker
halide	Halogenid	stir (in)	(hinein)rühren
precipitation	Niederschlag, Fällung	vigorous	kräftig

Procedure

The approximately neutral chloride solution is diluted in a 600 ml beaker to 150 ml.

After this we add 10 ml 2 n nitric acid free of chloride (10 ml concentrated acid and 40 ml distilled water).

Then we drop a 0.1 n nitrate of silver solution into this mixture until the precipitate starts flocculating. (This reaction shows a slight excess of Ag^+ ions). The process of conglomerating can be accelerated by adding 2–3 ml ether before the precipitation. As soon as the chloride of silver has settled, we have to convince ourselves whether the precipitation is complete or not by adding a few drops of nitrate of silver solution. We have to heat the solution slightly below the boiling point for a few minutes, while stirring it vigorously.

Now the liquid has to cool down in the dark. The solution should remain still for one night, at least for 2 or 3 hours.

Then we strain the liquid through a 1 G 4 filter crucible without stirring the precipitate. In order to prevent the precipitate from going colloidally into the solution while washing out, 0.01 n nitric acid is used as washing water. We have to wash out the precipitate three times with 20 ml washing liquid without pouring out the sediment. Then the precipitate has to be poured into the filter crucible by means of a rubber wiper.

Since the precipitate contains small quantities of the solution as a result of its cheeselike quality, it is more efficient to wash it out by decanting than later on in its conglomerate state. The washing out is continued in the filter crucible with small quantities of 0.01 n nitric acid until some ml of the filtrate show only slight opalescence after adding some drops of hydrochloric acid.

Finally we have to wash the sediment with a small quantity of pure water in order to remove the diluted nitric acid from the precipitate.

It is possible that the filtrate will turn a little turbid, because the AgCl is noticeably soluble in the washing water and precipitates again with the excess of Ag^+ ions in the filtrate.

Dry it at 130°C until it is constant.

Vocabulary

accelerate	beschleunigen	sediment	Niederschlag,
decant	dekantieren		(Boden)Satz
flocculate	ausflocken	strain	(durch)seihen,
pour out	ausgießen, -schütten		filtrieren
		turbid	trüb

Questions

1) What properties does silver chloride have?
2) Explain the procedure in your own words.
3) With what acid do you acidify?
4) When can you see that the precipitation is completed?
5) What are other possibilities of determining chloride ions?

Explain:

a) flocculent   b) colloidal(ly)   c) conglomerate
d) disintegrate   e) split

Grammar

Describing

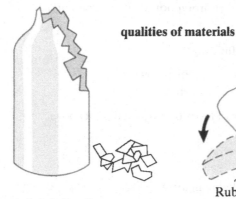

qualities of materials

Glass is brittle/fragile.
Glass is a brittle/fragile material.

Rubber is flexible/pliable.

Rubber is a ....

Make statements and describe the different properties of the following materials:

Materials	Properties	
wool	strong	fragile
polythene	tough	rigid
steel	resilient	weak
porcelain	flexible	soft
glass	stiff	flimsy
paper	elastic	hard
rubber	pliable	
wood	brittle	

Now, ask questions about the properties of various materials, answer them and use modifiers where necessary

– Is wool an extremely rigid material?
– No, it isn't. It's a very soft pliable material.

(quite, not very, extremely, rather, fairly, very, etc.)

Now compare materials.

Rubber is very tough. Paper is not very tough.
Rubber is **far tougher** than paper.

Cardboard is quite strong. Paper is not very strong.
Cardboard is **slightly stronger** than paper.

But notice what we say with these properties:

Glass is **considerably more brittle/considerably less resilient** than wood.

a) paper/flimsy/wood
b) copper/ductile/iron
c) rubber/rigid/steel
d) cardboard/stiff/paper
e) iron/malleable/wood
f) paper/strong/cardboard
g) porcelain/resilient/material/ plastic
h) wood/hard/cardboard
i) copper/good/conductor/lead
j) iron/poor/conductor/aluminium

Here are some properties of liquids and fluids:

viscous	thin	creamy	free-flowing
oily	runny	thick	sticky

Name some substances which have some of the properties in the list above.
    Milk is a free-flowing white liquid.

....

Some substances are between solid and liquid in form.
    gel (jelly)   (adj.: gelatinous)
    cream   (adj.: creamy)

Some solids may be found in the following forms:
    powder   (adj.: powdery)
    crystals   (adj.: crystalline)
    granules   (adj.: granular)
    filings
    chips
    flakes   (adj.: flaky/flocculent)
    shavings

                               (adj.: fine)
                               (adj.: coarse)

Now use the words above to describe the following substances as fully as possible.
    a) toothpaste   b) oil for a motor car
    c) glue   d) chalk
    e) butter   f) jam
    g) honey   h) sand
    i) instant coffee   j) salt

In the following table we form nouns and verbs from the adjectives.

Adjective	Verb	Noun	German
soft	soften	softness	
rough			
weak			
coarse			
tough			
hard			

But not all nouns can be formed in this way. Find the corresponding verbs, nouns, and paraphrases for the following adjectives:

Adjective	Verb	Noun	German
resilient	make sth resilient	resilience	
brittle			
smooth			
ductile			
strong			
rigid			
pliable			
elastic			
malleable			
flexible			

Thin Layer Chromatography

Paper chromatography is a versatile technique, but restricted in its use by the fact that separations can only be carried out on fibrous materials (e. g. cellulose), as other important chromatographic materials such as silica gel, oxide of aluminium and gel-filtration materials cannot be produced in sheet form. To avoid this difficulty thin layers of these substances are applied to a proper plate. It's common use to apply them to glass plates. But thin layers on plastic or metal foils, resisting solvents, are available too.

In many respects thin layer separations resemble paper separations but the larger choice of layer materials means that distribution, adsorption, gel-filtration and ion exchange separations can be carried out by this method. Another positive aspect of thin layers is the possibility of speeding up the reaction. It is not unusual to get excellent separations within 20 to 40 minutes. Even separations on cellulose thin layers can be carried out in a fraction of the time necessary for paper chromatography. One reason for getting better and faster separations is the fine structure of most thin layer substances.

As mentioned before, almost any material can be used as a thin layer: micro-granulated or micro-crystalline cellulose (distribution chromatography), silica gel (distribution and adsorption chromatography), and oxide of aluminium (adsorption chromatography).

What solvent you choose depends on the kind of substance to be separated and on the material on which the separation is to be carried out. Basically, the polarity of the substance to be separated must suit the polarity of the solvent. There's a simple method allowing a quick decision on a certain solvent:

Spot some drops of test solution at points on a thin layer. Then apply, by means of a fine capillary tube, some drops of another solvent to the spots, which will then spread circularly.

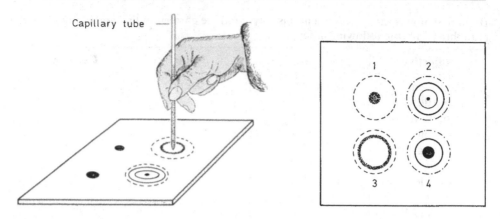

Capillary tube

In this case solvent no. 2 would be the most appropriate one.

To apply the test solutions to the origin line (as points or as lines) use a fine capillary tube or a micropipette. Allow the solvent, in which the substances are dissolved, to evaporate. As evaporation on thin layers takes place very fast, a hairdryer, or something of this kind, isn't necessary. You may apply different drops on the spot, as long as you allow the solvent to evaporate after each application.

Usually, the ascending method is used. Place the solvent at the bottom of the tank to a depth of 5 to 10 millimeters and stand the plate on the bottom. While developing, don't move the tightly-fitting lid. Once the chromatogram has run to the desired stage (usually a 10 cm rise), you take out the plate or foil and mark the solvent front with pencil. The solvent evaporates in a few minutes. If necessary, you can apply a little heat. Now it's possible to locate the test components.

If there is no commercial tank available, you can use ordinary powder-bottles or jam jars with screw tops.

Location of the spots is similar to that of paperchromatograms. One advantage of most thin layers is the fact that you can treat them with corrosive reagents (e. g., concentrated sulphuric acid) and strong oxidizing reagents. The reason for this is that they are purely inorganic by nature.

Explain in your own words what happens when fluorescent substances are added to thin layers and UV light is used.

1) tank (lid removed)
2) solvent
3) carrier plate
4) layer
5) solvent front
6) origins
7) substance with low sweep speed
8) substance with high sweep speed

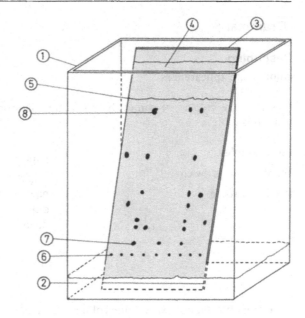

Vocabulary

appropriate	angemessen	granulated	körnig
ascend	aufsteigen	jam jar	Marmeladeglas
capillary tube	Kapillare	resemble	ähneln
carrier plate	Trägerplatte	screw top	Schraubdeckel
fibrous	faserig, Faser . . .	silica gel	Kieselgel
fluorescent	fluoreszierend, schillernd	thin layer	Dünnschicht . . .
		tight(ly)-fitting	genau an-, eingepaßt
foil	Folie, Unterlage	versatile	vielseitig

Questions

1) What kinds of chromatography do you know?
2) What are they used for?
3) Explain the procedure in your own words.

Paraphrase:

a) versatile
b) restrict
c) resemble
d) tightly-fitting
e) corrosive

| Grammar |

Describing
colours and **appearances**

Describe the colours of some objects in the classroom. Ask and answer questions using this table.

For example:

What colour is the chair?

It's dark brown.

light	blue
bright	red
pale	orange
dull	pink
deep	purple
	....

When an object is not exactly one colour, we can add **-ish** to the colour.

For example:

red	reddish
yellow	yellowish
blue	bluish

When an object is between two colours, we often say:

reddish-brown	bluish-yellow
greyish-green	etc.

It's also possible to say:

lightish blue	darkish grey

e.g.:   Copper is a reddish-brown colour.
         The sea is a bluish-green colour.

What colours are these?

amber	bronze	crimson
mauve	turquoise	khaki

Objects have different types of surface and appearance. They can be:
bright/shiny/dull/mat(t)/glossy

They can also be:
uneven / rough / coarse / grainy / pitted / corrugated / smooth / abrasive

Now put in the correct words in the following sentences:

The inside of a camera has a .... surface.

Glass is a .... solid which usually has a ....surface.

Mercury is a liquid metal which has a ....appearance.

Chalk is a porous solid which has a .... surface.

Some cardboard is .... to give it extra strength.

Carry on and form your own sentences.

Photometry in the Visible Range of the Spectrum

Fundamentals

The photometry in the visible range of the spectrum serves to solve quantitative analytical tasks, whereby it usually deals with the determination of elements, but also with that of molecules. A prerequisite is that the optical quantity to be measured is directly proportional to the concentration to be measured. First of all, diluted solutions are examined; the analysis of diluted gases is also possible. The measurements are always of a comparative nature. Solutions or gases of known composition, determined under the same conditions, serve as a reference system. It is possible to determine the substance you are looking for directly, by the evidence of its own optical behaviour, or indirectly, by the optical qualities of one of its coloured reaction products – be it complexes with an added auxiliary reagent or its own higher respectively lower valences. Usually, the latter is used; therefore the optical measurement is connected with a chemical reaction.

Laws

The necessary connection for photometric measurements between the optical quantity to be measured and the concentration of the substances to be determined is brought about by Lambert-Beer's Law. This states that there is an exponential proportionality between the intensity of the incident light and the intensity of the traversing light of the absorbing medium:

$$I_D = I_0 \cdot e^{-\varepsilon \cdot c \cdot d}$$

I_0 = light entering the sample
I = intensity after traversing the absorbing sample
ε = molar extinction coefficient
c = concentration of the absorbing sample
d = sample path length

The extinction coefficient is a function of the wavelength, but not interdependent with the concentration.

It's easier to use the Lambert-Beer Law in its decimal-logarithmic form:

$$\log \frac{I}{I_0} = -\varepsilon \cdot c \cdot d$$

The quotient I/I_0 is called transmittance (T), the reciprocal quotient I_0/I is often named absorption (A), and the logarithm I_0/I extinction (E) (in the case of natural logarithms – natural, in the case of decimal logarithms – decimal). It is recommended to work with extinction, because there is a linear relationship to the concentration of the analytical substance to be determined.

Lambert-Beer's law is based on monochromatic light only, i. e. light of a single wavelength (in practical work: light of a very narrow range of wavelength) and constant external conditions, such as temperature and solvent. Strictly speaking, it is a law for highly diluted solutions; its upper limit of validity is of the order of a 10^{-2} mol/l concentration. The condition of monochromacity must be fulfilled with greater precision if higher accuracy of measurement is desired.

Apparatus

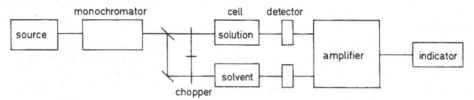

A spectrophotometer. The source and monochromator may be replaced by a tunable laser.

Normally, filament lamps are used for the examinations in the visible range of the spectrum. They must have a rather short luminous coil, and a high constancy. This means that the light sources must be operated with a very constant potential (the intensity of radiation of a filament lamp changes with the third to the fourth power of the working voltage; then the spectral distribution of the emitted light changes with the temperature of the incandescent filament of the lamp, which, in turn, depends on the working voltage). When working with light of a shorter wavelength, the intensity of the filament lamps decreases rapidly because of the low transmission of glass (it approaches UV). Light sources of this kind create a continuous spectrum and can only be used up to 300 nm.

As well as filament lamps, gas discharge lamps are being preferred because they emit only a few, but very intensive rays; thus they can easily be turned into monochromatic lamps. Often simple filters will do.

To create monochromatic light, monochromators are used. They can be prisms, gratings, and also filters.

The substance to be measured is transferred into cuvettes. They consist of thin, planeparallel glass – and quartz plates of extreme clarity. These conditions need not be met by the sides. Therefore they may consist of less valuable glass or quartz.

For measurements of liquids the optical paths are between 0.1 cm (micro-cuvettes) and 5 cm. When examining volatile liquids (for example, as solvents), it is necessary to cover the cuvettes. Sometimes gas cuvettes have considerably bigger optical paths. Glass cuvettes can be used in the range of wavelengths between 350 and 800 nm, those made of quartz between 200 and 1000 nm (but in practical work the latter only if the other optical paths of the measuring instrument are made of quartz, too.)

Procedure

The procedure of photometric measurements is fairly simple. After having reached the working constancy of the measuring instrument, the solvent is first transferred into the cuvette; while measuring the extinction, the apparatus is adjusted to the value '0'. Then the cuvette, filled with the solution to be measured, is photometrified. The value thus obtained is converted to the amount of the substance present by means of calibration curves.

If possible, select a measuring wavelength which corresponds to the extinction maximum of the substance to be measured. In this way the highest sensitivity is reached.

The calibration curve is taken up with a known quantity of the substance to be measured in the way that increasing quantities of it are transferred into the same volume in which the analytical sample is present. The pH-value of the solution and the salt content are to be the same as those of the analytical sample.

Calibration measurements and analytical measurements must take place at the same temperature, usually room temperature, as the extinction coefficient is dependent on temperature.

If the coloured component only has a low concentration, it is recommendable to work with cuvettes with big optical paths. More concentrated solutions should be measured in small optical paths.

Vocabulary

adjust to	einstellen, justieren	prerequisite	Voraussetzung
chopper	Zerhacker	reciprocal	reziprok
coil	Wicklung, Spule	recommend-able	empfehlenswert
emit	aussenden		
fairly	ziemlich	reference system	Bezugssystem
filament	(Glüh-, Heiz-)Faden		
incident	einfallend	strictly speaking	streng/genau genommen
inter-dependent	voneinander abhängig, eng zusammenhängend	transmittance	Durchlaßgrad
		traverse	hindurchgehen
luminous	leuchtend, Leucht. . .	valence	Wertigkeit
optical path	Schichtdicke	validity	Gültigkeit

Questions

1) Show in diagram form the order in which the apparatus for photometry should be used.
2) What sources of light and what different means besides the filter can be used?
3) What do we call the vessel which contains the sample?
4) Describe the procedure.
5) What fundamental law concerning photometry do you know?

Elucidate:

a) prerequisite   b) incident light   c) traversing light
d) monochromatic light   e) accuracy   f) recommendable

Describing

an activity

Here are two diagrams showing how the air in a room is heated by **convection.**

The first shows diagrammatically what happens inside the room, while the second sets out the sequence from which convection currents arise.

Study both of the diagrams and use them to account for the phenomenon of convection, explaining what causes air to rise, fall, expand, etc. Ask and answer questions about the sequence.

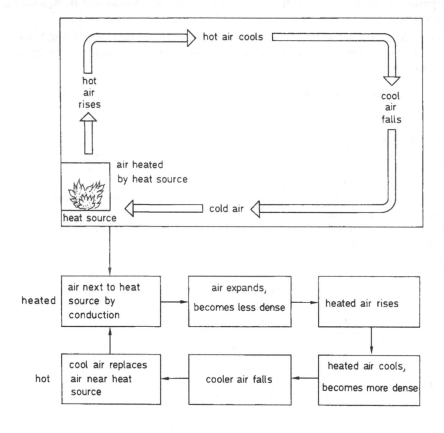

When matter changes its state, it does so in stages. Describe the stages of change when **ice melts,** when **sulphur is heated,** when **water is used in an electricity generator,** etc.

The following words will be useful:

melt	solidify	freeze
vaporize	condense	evaporate
liquefy	sublimate	

Use the correct verbs in these statements:

When a substance changes from a solid to a liquid, it is said to ....

At normal temperatures, iron is solid. However, when it is .... above

1.537 °C, it ....

When a substance changes from a liquid to a gas, it is said to ....

When water is .... to 100°C, it ....

Water is a liquid at normal temperatures. However, when it is ....

below 0°C, it ....

When a substance changes from a liquid to a solid, it is said to ....

Conductivity Meter

Course of measurement

The unknown resistance respectively the measuring cell for conductivity is connected to the terminal 'Cell'. The mains switch is turned from 'Off' to 'On' after having connected the instrument to mains.

After 1 to 2 minutes the instrument is ready for operation. It is expedient to select the measuring frequency indicated at the range switch for the relative range:

high conductances 3 kc/s; small conductances 50 c/s.

If the value is to be determined in ohms or mhos the left switch has to be turned to 'Ohm' or 'Mho' respectively. The range switch with black factors applies for ohms; for mhos the switch with the red factors applies. The compensator for the loss angle is turned to zero (angle centre).

The range switch is turned to the range expected. Now the centre pointer knob is turned until the indicating instrument shows the minimum. It is expedient to place the pointer at minimum and thus encircle the balance position. (With some experience it is easy to notice even the smallest deviations). Then the scale value is read off and multiplied by the factor indicated on the range switch. A double line on the pointer on

WTW-Konduktometer[12]

the reading scale provides a reading free from parallax. If, during the minimum determination, the pointer of the indication instrument does not go back to the scale beginning, a zero position as near as possible to the scale beginning is to be found. This is to be done by alternatively adjusting the loss angle adjuster and the measuring indicator.

Care has to be taken that the solution to be measured is not in connection with the soil. Otherwise a measurement is impossible.

The following points are important when selecting the measuring frequency (50 or 3000 c/s):

3000 c/s are used where polarisation of the substance may have a disturbing influence, i. e. in the event of higher conductances;
50 c/s are suitable in cases where the capacitance on the conductance to the measuring cell may have disturbing effects in the event of small and the smallest conductivities.

> **Note:** If measurements are to be carried out in grounded substances such as rivers, containers, pipelines, etc., it is necessary to avoid conducting connection to the soil. For this purpose the 4 chromed screws on the frontal plate are to be unscrewed. The instrument is pulled out with the two handles. The grounding vein (from power supply cable) is to be separated from the chassis and insulated. Re-insert the instrument and screw the screws in place. (Not necessary for battery operation).

Attention! Housing is no longer earthed!

When using conductivity the cell constant has to be taken into account. This constant (k), divided by the resistance measured (ohm), is the conductivity value in question. If the mho-scale is used, the cell constant (k) has to be multiplied by the mho-value **measured.** These figures are also stated on the scale of the instrument.

$$\varkappa = \frac{k}{R\ (\text{ohm})}$$

$$\varkappa = k \times S \qquad\qquad (\text{Siemens} = \text{mhos})$$

$\varkappa$ = spec. conductivity in $\text{ohm}^{-1}\text{cm}^{-1}$, resp. S (mhos) cm^{-1}
k = cell constant in cm^{-1}
R = determined resistance in ohm
S = determined reciprocal resistance in ohm^{-1} resp. Siemens (ohms)

If measurements are not performed at a constant temperature (connection to a thermostat) the measuring value is to be corrected with the temperature coefficient referring to the reference temperature (e. g. 20°C). The temperature coefficient is, according to substance, about 1.5–2.3% per °C. The temperature coefficient is to be deducted in the event of excess temperature and to be added in insufficient temperature.

Vocabulary

approach	sich nähern; nahekommen	grounding vein insufficient	Massekabel unzulänglich,
capacitance	Kondensator-kapazität		unzureichend, ungenügend
conductance	Leitfähigkeit	insulate	isolieren
deduct	abziehen, ein-behalten	mains switch parallax	Hauptschalter Parallaxe
deviation	Abweichung; Ablen-kung	range resistance	Bereich Widerstand
earthed	geerdet	take into	berücksichtigen
encircle	einkreisen	account	
expedient	ratsam, zweckmäßig	unscrew	ab-, auf-,
grounded	geerdet		losschrauben

Questions

1) What solutions conduct electricity?
2) What kind of current is used?
3) What happens when direct current is applied?
4) Do you determine the concentration of a solution by just one measurement?
5) Draw an example of a conductrimetrical titration curve.
6) Draw the schematic arrangement of a conductometer. Explain.
7) What's the main difference between conductivity and potentiometry?

Paraphrase:

a) deviation b) adjust c) take into account
d) deduct e) insufficient

Describing

a process

The Process of Generating Electricity

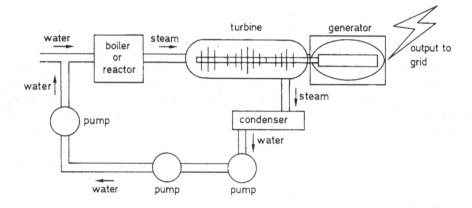

Steam is produced in either a boiler or a nuclear reactor. In the case of a boiler, this may be fuelled by either coal or oil.

The steam travels along pipes to a turbine, where it drives the shaft at high speed. The shaft of the turbine is coupled to the rotor of the generator, and the rapid revolution of the rotor induces an electric current in the outer part of the generator, which is known as the stator. This electricity is then fed into the electricity grid system.

When it has passed through the turbine, the steam enters the condenser. Here it is passed over tubes containing cooling water. The steam is therefore cooled, and it condenses back to water. The water is then returned to the boiler by means of a series of pumps.

Study this description carefully.

An Electrochemical Cell

Electrolysis is the separation of a substance into its chemical parts by electric current. Now, is it possible to construct such a cell in which, instead of a current decomposing chemicals, chemicals react to produce an electric current? Towards the end of the eighteenth century, Luigi Galvani and Alessandro Volta acquired fame by their discovery that this is indeed possible. Volta found that electricity was produced by a chemical reaction between two different metals, and made the first practical battery.

The two metals used by Volta in his battery were copper and zinc. It can be shown by experiment that if copper is added to zinc(II) sulphate(VI) solution no reaction takes place, whereas if zinc is added to copper(II) sulphate(VI) solution the blue colour of the hydrated copper(II) ion gradually disappears and a reddish brown sludge of copper is deposited.

Equations

$$Cu_{(s)} + Zn^{2+}SO_{4(aq)}^{2-} \rightarrow \text{no reaction}$$

$$Zn_{(s)} + Cu^{2+}SO_{4(aq)}^{2-} \rightarrow Cu_{(s)} + Zn^{2+}SO_{4(aq)}^{2-}$$

The reaction above can be explained in terms of electron loss and gain. Each atom of zinc loses two electrons to each copper ion in the solution:

$$Zn - 2e \rightarrow Zn^{2+}$$

$$Cu^{2+} + 2e \rightarrow Cu$$

These are called **half-reactions.** The half-reaction in which electrons are lost is called **oxidation,** and the half-reaction in which electrons are gained is called **reduction.** Although there is a transfer of electrons, an electric current is not produced unless the electrons are able to flow from the zinc **through a circuit** (e. g. a wire) to the copper(II) ions in solution. A chemical device for creating a flow of electrons in a circuit is called a **primary cell.**

The two electrolytes must be in contact and yet not allowed to mix. The apparatus shown on the next page uses a cardboard partition which, on becoming soaked, provides electrical contact without mixing of the electrolytes. However, cardboard is not very strong and a more satisfactory partition is the porous pot used in the **Daniell cell.**

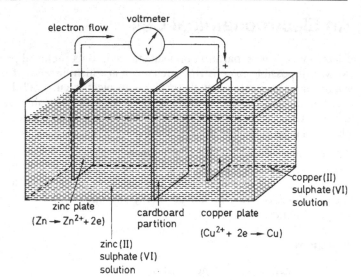

A simple primary cell

The zinc rod in the Daniell cell gradually dissolves as atoms of zinc lose electrons to become zinc(II) ions. These electrons pass through an external wire circuit to the copper container where copper(II) ions from the copper(II) sulphate(VI) solution are converted into metallic copper. Copper(II) sulphate(VI)crystals are present to maintain a saturated solution. A voltmeter connected across the terminals of the cell registers 1·1 volts. This voltage, recorded when there is no current flowing, is known as the **electromotive force** (e. m. f.) of the cell.

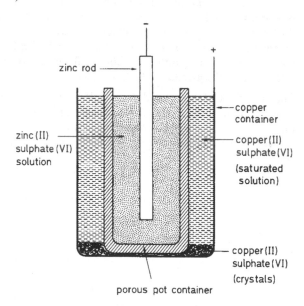

The Daniell cell

By using different combinations of metals in a suitable electrolyte (often a solution of the metal salt), it is possible to construct a variety of electrochemical cells delivering different voltages.

Vocabulary

acquire	gewinnen, erhalten	partition	Scheidewand
cardboard	Pappe	porous	porös
circuit	Kreis	provide	liefern, versorgen
current	Strom	saturated	gesättigt
decompose	(sich) zersetzen; sich	sludge	Schlamm, Bodensatz;
	auflösen, zerfallen		Klärschlamm
deliver	abgeben, liefern	soak	sich vollsaugen,
deposit	(sich) ab-, nieder-		durchtränkt werden;
	setzen, ablagern; sich		einweichen
	niederschlagen	terminal	(Anschluß),
device	Gerät, Vor-, Ein-		Klemme, Pol
	richtung	unless	wenn . . . nicht;
electromotive	elektromotorisch		außer wenn
fame	Ruhm	variety	Anzahl, Reihe; Man-
hydrated	wasserhaltig		nigfaltigkeit; Vielfalt
in terms of	in Form von, im	voltage	Spannung
	Sinne, hinsichtlich,		
	vom Standpunkt		

Questions

1) What is 'electrolysis'?
2) What did Luigi Galvani and Allessandro Volta discover?
3) What do we understand by 'oxidation' and 'reduction'?

Find synonyms for:

a) decompose   b) acquire   c) fame
d) gradual(ly)   e) suitable   f) deliver

Exercise

Translate

Electrolysis

Electrolysis is defined as a chemical change brought about by the use of an electric current source. Electrolysis can take place in an electrolytic cell (diagrammed below). The electrolytic cell includes two **electrodes** which are connected to a source of electricity and an **electrolyte** which can conduct electricity when in liquid form or in solution. The electrolyte in the diagrammed electrolytic cell is molten NaCl. The term **molten** indicates that solid NaCl has been melted to a liquid state.

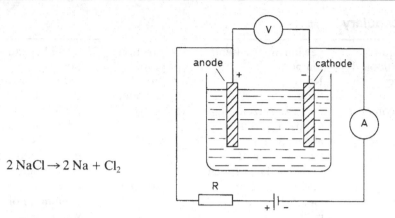

$$2\,NaCl \rightarrow 2\,Na + Cl_2$$

Atomic Structure

An atom, the smallest unit of a chemical element, is composed primarily of three fundamental particles: electrons, protons, and neutrons. The combination of these particles in an atom is distinct for each element.

Unless otherwise stated, an atom is assumed to be **neutral.** In any neutral atom, the number of electrons is always equal to the number of protons.

– A boron atom contains five protons. We assume the atom to be neutral. How many electrons must it have?

Bohr's Atomic Model

A Danish physicist, Niels Bohr, came up with a model which pictured the atom with a nucleus of protons in the center and electrons spinning in an orbit around it (similar to the movement of the planets around the sun). The following Bohr model contains one orbiting electron and a nucleus of one proton.

– What is the atomic number of the element represented?
– What element is represented?

Two Frequently Applied Techniques of Separation

Filtration

We expect our car to perform satisfactorily under all conditions, and we take it for granted that the water we drink is pure. Solid material has been removed from engine oil, dirt from the air used in the car engine, and solid particles from drinking water by a process called **filtration**. This is a method of separating a solid from a liquid (or a solid from a gas) by passing the mixture through a **filter**. The filter consists of a porous material (such as paper or glass wool or fine gravel) which prevents solid particles from passing through but does not retain the liquid.

Experiment 1: Separation of salt and sand

A teaspoonful impure rock salt is stirred into a beaker of warm water. The suspension is filtered into a beaker or test tube. The clear colourless solution (called the **filtrate**) obtained in the test tube contains the salt; the sand and other impurities remain behind on the filter paper.

Distillation

Distillation can be used to separate a **volatile** substance (i. e. one which evaporates fairly readily) from a non-volatile substance. It consists of the two basic processes of boiling and condensation.

Experiment 2:

Impure water can be purified by distillation because the ordinary impurities in water are solids which are non-volatile. The impure water is heated to make it boil; its vapour is collected and cooled. As it cools, the vapour **condenses** into pure **distilled water**. The non-volatile solid impurities remain behind.

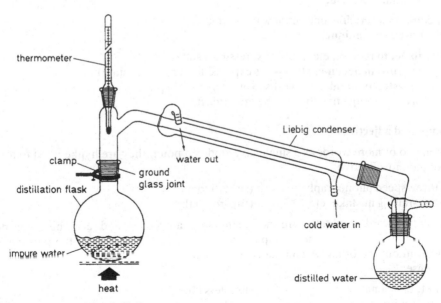

Laboratory apparatus for distillation

Vocabulary

gravel	Kies	take for	als selbstverständlich
impure	unrein	granted	betrachten,
perform	(durch-, aus-)führen,		-hinnehmen
	funktionieren	volatile	(leicht)flüchtig
retain	zurück(be)halten		

Questions

1) What is the purpose of filtration?
2) What does the word 'suspension' mean?
3) By what means can a filtration be carried out?
4) What do we understand by 'distillation'?
5) What sorts of substances can be distilled?
6) What happens in the distillation of a salt solution?

Explain or give synonyms for:

a) satisfactory   b) take something for granted   c) porous

d) retain   e) impurity   f) volatile

Grammar

Adjective

An adjective is a word that modifies a noun.

Look at the following two sentences and explain what we understand by **attributive use** and **predicative use.**

Air contains a **variable** amount of water vapour.
This substance is **unique.**

After **to be, to remain,** etc. (verbs expressing a state)
    to grow, to become, etc. (verbs expressing a change of state)
    to taste, to sound, etc. (verbs expressing a perception)
use a **predicate adjective** to describe the subject.

Compound adjectives

When two or more words as a single unit modify a noun, they are hyphenated to form a compound adjective:

    three-speed phonograph   (six-cylinder car)
    hair-raising mistake, etc.   (two-room flat)

Many two-word modifiers are not compound adjectives and are therefore not hyphenated. Here are some examples. Notice that each adjective of the pair makes sense without the other. A comma is used if **and** could be inserted without changing the sense:

    hot, dry climate   cloudless blue sky
    strong, firm clasp   neat little craft, etc.

Comparison

Most adjectives of one syllable and some adjectives of two syllables form the comparative and the superlative by adding **er** and **est** to the positive:

clear	clearer	clearest
lazy	lazier	laziest

Adjectives of three or more syllables are compared by putting **more** and **most** (or **less** and **least**) before the positive:

durable	more durable	most durable
beautiful	more beautiful	most beautiful, etc.

Many adjectives of two syllables form the comparative and the superlative by putting **more** and **most** (or **less** and **least**) before the adjective:

careless	more careless	most careless
futile	more futile	most futile, etc.

Several adjectives are compared irregularly: (Complete.)

bad, evil, ill	worse	worst	
good	....	....	
much, many	....	....	
little	....	....	– denoting quantity
little	....	....	– denoting size
far	....	....	– denoting distance
far	....	....	– denoting what comes in addition ('weitere Beiträge')
late	....	....	– denoting a relation of time (**the latest issue:** die neueste Ausgabe)
	the latter	the last	– denoting sequence (the latter = der letztere)
old	....	....	– denoting opposite of young
old	....	....	– denoting family relationship (attributively) (her eldest brother)
near	....	....	– denoting distance
		next	– denoting sequence (the next analysis)

Exercises:

A Pair each adjective in columns 1 and 2 with an appropriate noun in columns 3 and 4, and give the comparative and superlative of the adjective with its noun:

fast	fascinating	student	explosion
bad	rapid	biography	reaction
quiet	clear	colour	night
little	terrifying	chemist	quantity

B Translate into German

most people / the most beautiful flowers / most of the animals / most complicated techniques / the most complicated techniques / the most common computers in use today / the most commonly used type.

---

o-Anisaldehyde

Procedure

A *4-Dimethylamino-2'-methoxybenzhydrol*

An ether solution of *o*-methoxyphenylmagnesium bromide is prepared in the usual manner with 250 ml. of anhydrous ether, 14.5 g. (0.60 g. atom) of magnesium, and 100 g. (0.53 mole) of *o*-bromoanisole. A solution of 60 g. (0.40 mole) of *p*-dimethyl-aminobenzaldehyde in 200 ml. of anhydrous benzene is added drop by drop. After the addition is completed, the reaction mixture is stirred for 10 hours at room temperature. The magnesium complex, which forms a very thick suspension, is decomposed with a solution of 75 g. of ammonium chloride in 450 ml. of water. The ether-benzene layer is separated, washed with 100 ml. of water, and dried over calcium sulphate. The solvent is removed under reduced pressure, and the residue is induced to crystallize by trituration with a little petroleum ether (30–60°). Recrystallization of the solid from benzene-petroleum ether (30–60°) gives 4-dimethylamino-2'-methoxybenzhydrol (59–60 g., 57–58%), m. p. 75–80°.

o-Anisaldehyde

B   *o-Anisaldehyde*

In a 3-l. three-necked flask fitted with a mechanical stirrer and a nitrogen inlet tube are placed 60 g. (0.35 mole) sulfanilic acid, 18 g. (0.17 mole) of anhydrous sodium carbonate, and 400 ml. of water. Stirring is started, and the resulting solution is cooled to 0–5°C in a ice bath. Nitrogen is passed into the reaction flask, and a nitrogen atmosphere is maintained throughout the reaction. To the cooled solution is added three-quarters of a solution of 24.2 g. (0.35 mole) of sodium nitrite in 75 ml. of water, followed by 32 ml. of concentrated hydrochloric acid. During diazotization the temperature of the solution is maintained below 5°C by the action of ice in small pieces.

After a few minutes another 32 ml. of acid is added. Further additions of the sodium nitrite solution are made slowly until a positive test for excess nitrous acid is observed. The diazonium solution is buffered to pH ~ 6 by the addition of a cooled solution of 50 g. of sodium acetate in 125 ml. of water. A solution of 52 g. (0.20 mole) of 4-dimethylamino-2'-methoxybenzhydrol in 500 ml. of acetone is added rapidly, followed by an additional 500 ml. of acetone. The reaction mixture becomes red almost immediately, and stirring is continued for 30 minutes at 0–5°C. The cooling bath is replaced by a warm water bath (50–60°C), and stirring is continued for an additional 30 minutes. The reaction mixture is diluted with an equal volume of water and extracted with three 750-ml. portions of ether. The combined ethereal extracts are washed with water until all the dissolved methyl orange is removed, then dried over calcium sulphate.

The ether is removed under reduced pressure, and the residue is distilled to yield 19–20.5 g. (69–75%) of colourless liquid, b. p. 79–80° C (1.5 mm.) n^{25}D 1.5586.

Methods and merits of preparation

o-Anisaldehyde is commercially available. However, this procedure illustrates a method of general applicability for the preparation of aromatic, aliphatic, and unsaturated aldehydes and ketones.

Vocabulary

anhydrous	wasserfrei	induce	veranlassen; verursachen; auslösen; ableiten
applicability	Anwendbarkeit, Eignung		
buffer	(ab)puffern	inlet tube	Einlaßrohr, Einflußröhre
diazotization	Diazotierung		
ethereal	etherartig	ketone	Keton
fit with	ausrüsten mit	residue	Rückstand
		triturate	zerreiben

Quelle: Org. Synth. Coll. Vol. V, 46, 1983.

Questions

1) Why does the ether have to be anhydrous?
2) What apparatus, instruments etc. do we need?
   What are their functions?
3) Give a short description of the procedure.

Elucidate:

a) suspension   b) residue   c) extract
d) available   e) applicability

Grammar

Adverb

Have a look at the following sentences and state what the adverbs modify:

1) The fire burned **slowly:**   ....

2) This method of analysis is **very** inadequate: ....

3) He carried out this experiment **only** once: ....

4) **Unfortunately** the assistant chose the
   wrong substance.   ....

Adverbs answer the questions:			**Type of adverbial**
"When?"	(Yesterday)	⟶	adverbial of **time**
"Where?"	(.................)	⟶	....
"How?"	(.................)	⟶	....
"How much?"	(.................)	⟶	....
"How often?"	(.................)	⟶	....

It is usually easy to recognize an adverb that has been formed by adding **ly** to an adjective:

rapid   – rapidly
vigorous   – vigorously
gentle   – gently, etc.

Ly, however, is not a sure sign, for many adjectives like
    silly, holy, curly, etc.
end in **ly.**

The following words which do not end in **ly** are commonly adverbs:

afterwards	ever	now	seldom	there
almost	forward	often	so	too
already	here	once	sometimes	twice
also	just	perhaps	somewhat	very
altogether	never	quite	soon	well
always	not	rather	then	yesterday

Comparison

Adverbs ending in **ly** form the comparative and superlative by putting **more** and **most** before the positive:

gently   more gently   most gently

Adverbs which do not end in **ly** commonly add **er** and **est** for the comparative and superlative:

soon	sooner	soonest
fast	faster	fastest

Several adverbs are compared irregularly: (Complete.)

well   better   best

badly, ill   ....

little   ....

much   ....

Position

Insert the adverbials into their correct position in the sentence and state the rules:

1) Electricity can be produced. (also, very cheaply, in Norway, by water power)
2) The heated mercury was converted into a red powder. (slowly)
3) Sodium attacks water. (violently)
4) 17 ml of the acid are added. (quickly, to the alkali, then)
5) Sodium attacks water. (always)

Exercises

A Form an adverb from each of the following words, and use it in a sentence illustrating its meaning and use.

1) considerable   2) gravimetric   3) experiment   4) public
5) electric   6) initial   7) tremendous   8) natural

B Which of the words ending in **ly** are adverbs? Which are adjectives?

1) lively conversation   2) quickly accomplished   3) surly answer
4) only Mother   5) early bird   6) really happy
7) fully aware   8) burly man   9) vigorously stirred

Ethanol

Of all the alcohols the most important is **ethanol,** C_2H_5OH. Its presence in intoxicating beverages is well known: beers and wines contain up to 10% of ethanol, whereas spirits such as whisky contain about 40% by volume. Pure ethanol (**absolute ethanol**) is described as '200° proof', so a drink labelled '80° proof' contains 40% ethanol by volume. However, most of the ethanol manufactured is used industrially as a starting material for other chemicals such as ethanoic acid (acetic acid).

The two most important methods used for the manufacture of ethanol are

- the **fermentation of sugar or starch,** and
- the **hydration of ethene**

from petroleum.

Fermentation

Fermentation is a chemical action brought about by bacteria or yeasts. Living yeast produces biological catalysts called **enzymes.** When yeast is added to a dilute solution of ordinary table sugar a reaction occurs which proceeds most readily at about 38°C. The enzyme **sucrase,** produced by the yeast, catalytically breaks down ordinary table sugar (sucrose, $C_{12}H_{22}O_{11}$) into simpler sugars **glucose** and **fructose.** These are isomers having the molecular formula $C_6H_{12}O_6$. **Zymase,** a second enzyme produced by the yeast, then converts the glucose and fructose into ethanol and carbon dioxide.

$$C_{12}H_{22}O_{11(aq)} \; + \; H_2O_{(l)} \; \xrightarrow{\text{sucrase}} \; C_6H_{12}O_{6(aq)} \; + \; C_6H_{12}O_{6(aq)}$$
$$\text{(sucrose)} \qquad\qquad\qquad\qquad\qquad\qquad \text{(glucose)} \qquad \text{(fructose)}$$

$$C_6H_{12}O_{6(aq)} \; \xrightarrow{\text{zymase}} \; 2\,C_2H_5OH_{(aq)} \; + \; 2\,CO_{2(g)}$$
$$\text{(glucose or fructose)} \qquad\qquad \text{(ethanol)}$$

When the reaction mixture contains about 12% by volume of ethanol the activity of the yeast ceases. The ethanol can then be concentrated by fractional distillation if required.

Ethanol from Petroleum

Large quantities of synthetic ethanol are manufactured from ethene, a gas produced by the cracking of petroleum. The ethene is **hydrated** (a molecule of water is added) to produce ethanol:

Sulphuric(VI) acid assists in the hydration process.

Vocabulary

beverage	Getränk	intoxicate	berauschen
crack	(zer)spalten	label	bezeichnen; mit einer
ethene	Äthylen, Ethylen		Aufschrift versehen
fructose	Fruchtzucker, Fruk-	starch	Stärke
	tose	sucrase	Saccharase,
glucose	Stärkezucker,		Glukosidase
	Traubenzucker	yeast	Hefe
hydration	Wasseranlagerung	zymase	Zymase

Questions

1) Give some examples that show the importance of ethanol.
2) Describe alcoholic fermentation.
3) How do you obtain ethanol from ethene?

Explain:

a) intoxicating beverage   b) label   c) cease

Exercises

A   Translate and discuss

Plastics

A short look at the history of plastics

The use of plastics began slowly in the last half of the 19th century following the introduction of **celluloid** and other cellulose plastics (1869). The invention of **bakelite** (1907/09) brought about a marked increase in the use of plastics early in the 20th century. Shortages of natural materials which were the results of World War I and II led to the development of new plastic substitutes.

By the 1960s plastics were available in forms that were superior, for certain uses at least, to leather, wood, metal, paper, glass, rubber, and other natural materials. Plastics were used for a wide variety of purposes; for example, in the manufacture of automobile parts and bodies, clothing, household goods, toys, cigarette packages, television, etc.

The modern plastics industry deals mainly with mouldable materials manufactured from organic compounds of high molecular weight, which are known technically as macromolecules or polymers.

Chemically, the basic resins consist of combinations of carbon with hydrogen, oxygen, nitrogen, chlorine, fluorine, silicon, and other elements. The resins usually are compounded with plasticizers, stabilizers, fillers, dyes, and pigments to impart properties desired for specific commercial uses and for processing.

The three major ways of processing plastics are:
pressing – injection moulding – rolling

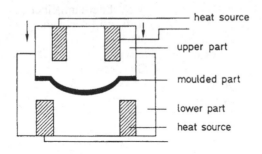

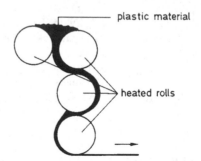

Explain the two procedures shown above.

The world-production of plastics:

1900: 20 000 t.	1913: 50 000 t.	1939: 0.35 mill. t.
1947: 0.87 mill. t.	1950: 1.5 mill. t.	1955: 3.3 mill. t.
1960: 7 mill. t.	1969: 24 mill. t.	1980:

Comment on the above figures and give reasons why there has been such a gigantic increase in the production of plastics. Use also the terms:
    low specific gravity;
    corrosion-resistant;
    easily moulded.

Questions

1) Describe plastics in the simplest terms possible.
2) What resins are plastics based on?
3) What can be said about the properties of plastics?
4) What articles are made of plastics?

Theme

Besides many advantages, plastics have disadvantages, too. Name some:
– plastic waste;

B   Read and translate into German:

Responding generally to the environmental protection movement and to the growing shortage of petrochemicals from which most plastics and polymers are manufactured, a trend toward recycling of discarded plastics picked up momentum in 1976. Plastic wastes are pyrolyzed or broken down at high temperatures in an inert atmosphere into useful hydrocarbons and other products. Depending on the plastic starting material and the temperature at which it is pyrolyzed, the reaction products can vary widely. Plastic waste, if recycled efficiently, should relieve some of the strain on world resources of oil and help solve an environmental problem by eliminating the vast amounts of plastic waste accumulating upon the earth.

C Translate into English:

Vor einigen Jahrzehnten wurden Kunststoffe noch als minderwertige Ersatzstoffe angesehen. Diese Ansicht hat sich jedoch grundlegend geändert, seit man erkannte, daß viele Kunststoffe weit bessere Eigenschaften haben als manche der bisher benutzten Werkstoffe. Kunststoffe sind nicht nur neuartige, sondern auch neuwertige Werkstoffe, die oft eine so große Zahl von guten Eigenschaften in sich vereinigen wie kein natürlicher Werkstoff. Ein Beweis, wie hoch Kunststoffe im täglichen Leben und in der Technik bewertet werden, ist die Zunahme der Produktion.

Vocabulary

body	Körper; Karosserie; Rumpf (Schiff)	mould	formen, gestalten; gießen, pressen
cast	gießen	mould	Form, Preßform, Preßwerkzeug
coating	Überzug, Belag; Beschichtung	moulding	Formen, Formpressen, Preßverfahren; Preßteil
compound	Verbindung		
compound	zus.setzen, verbinden; (zu)bereiten, herstellen	plastic	Kunststoff, Plast, Plastik; Preßstoff, Kunstharz
convert	umformen, umändern, umwandeln	plastic (adj)	plastisch, bildsam; (ver)formbar; aus Kunststoff
dye	Farbstoff, Farbe		
filler	Füllmasse, Füllstoff; Streckmittel; Abfüllmaschine	plasticizer	Weichmacher, Plastifiziermittel
		processing	Verarbeitung, Bearbeitung; Veredelung
houseware (US)	Haushaltswaren		
household goods (Br.)		property	Eigenschaft
		resin	Harz, Kunstharz
injection	Einspritzung	roll	Walze, Rolle
injection moulding	Spritzguß(verfahren)	specific gravity	spezifisches Gewicht
		substitute	Ersatz(stoff), Austausch(werk)stoff, Surrogat
milling	Mahlen, Zerkleinern; Walzen; Fräsen		

3 Chemistry – In Everyday Life

With 5 000 000 000 Passengers . . .
High Priority
Recycling
Pesticides, Fungicides, Herbicides
Chemistry is when . . .
Pollution: As Old As Mankind Itself

With 5 000 000 000 Passengers...

With 5 000 000 000 passengers it would be surprising if spaceship Earth could provide a good diet, a comfortable cabin and a useful occupation for all.

With another 2 000 000 000 due to come aboard in the closing years of the century, even the cleverest captain would be a little anxious about accommodating them, particularly as certain supplies are running short and as the first-class passengers are so greedy.

But our planet does not have a captain. Power lies with scattered groups of crewmen and some better-off passengers who generally put their own interests first. Many of the other passengers are living in abominable conditions. And all are reliant on the fragile biosphere for survival. Someone should be monitoring their efforts, reporting on the damage they may do in their struggle for food, shelter and riches, and on their success in meeting needs without harming the natural world of which they are a part.

Population Growth

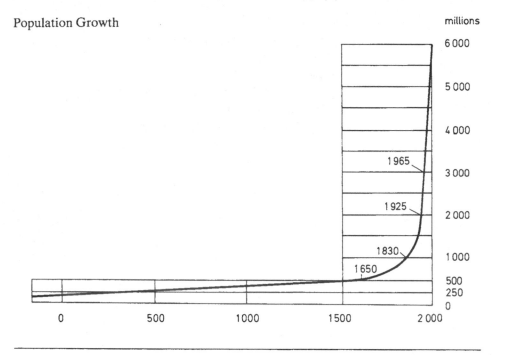

Vocabulary

abominable	abscheulich, äußerst übel	diet	Nahrung, Ernährung
		fragile	schwach; zerbrech- lich
accommodate	versorgen; unter- bringen		
		greedy	gierig
better-off	wohlhabender, besser dran	monitor	überwachen
		occupation	Beschäftigung

reliant on	abhängig von		scatter	(aus-, ver-,
run short	zur Neige gehen,			zer)streuen
	knapp werden		supplies	Vorrat; Versorgung

Questions

1) Why should the earth be likened to a space ship?
2) Explain what is meant by '. . . as the first class passengers are so greedy.'
3) What are the dangers facing the passengers? What counter measures should they be taking?

Paraphrase:

a) diet
b) occupation
c) accommodate
d) scatter
e) better-off
f) abominable
g) abominable conditions
h) fragile biosphere
i) monitor

Grammar

Prepositions

Exercise   Fill in the missing prepositions and form sentences:

to transfer .... überführen in

to result .... führen zu, bewirken

to deal .... handeln von, sich beschäftigen mit

to replace .... ersetzen durch

to succeed .... Erfolg haben bei, gelingen

to depend .... abhängen von

to be interested .... sich interessieren für

to consist .... bestehen aus

to insist .... bestehen auf

to rely .... sich verlassen auf

to separate .... trennen von

to differ .... sich unterscheiden von

to adhere .... haften an

to pay attention .... achten auf

to take part .... teilnehmen an

to limit .... beschränken auf

to confine .... beschränken auf

to apply .... gelten für

made .... steel   aus Stahl gefertigt

.... this reason   aus diesem Grunde

.... this way   auf diese Weise

.... the spot   auf der Stelle

.... 700 degrees   bis auf 700 Grad

.... want of   aus Mangel an

.... magnetising the rod   beim Magnetisieren des Stabes

.... this occasion   bei dieser Gelegenheit

.... Sunday   bis Sonntag

The most important and most frequently used compound prepositions are:

according to	gemäß, nach, laut
as to, as for	was . . . betrifft, bezüglich
because of	wegen, infolge von
by means of	mittels, durch
by reason of	wegen, infolge
by virtue of	kraft, auf Grund von, vermöge
by way of	durch, (auf dem Weg) über, in der Absicht zu, um . . . zu
due to	wegen, infolge
for the sake of	um . . . willen, wegen
in accordance with	in Übereinstimmung mit, gemäß
in addition to	außer, zuzüglich
in consequence of	infolge von, wegen
in front of	vor, gegenüber
in relation to	in bezug auf
in spite of	trotz, ungeachtet

in respect of/to	hinsichtlich, bezüglich, in Anbetracht
in view of	im Hinblick auf, angesichts
in virtue of	auf Grund von, kraft, vermöge
instead of	(an)statt, an Stelle von
on account of	um . . . willen, wegen
out of	aus (. . . heraus), zu . . . hinaus, von (two out of three – zwei von drei Personen etc.), außerhalb, außer (Reichweite, Sicht etc.), außer (Übung, Atem etc.), ohne (Geld etc.), aus der Mode, Richtung, nicht gemäß etc., außerhalb (5 miles out of London), (hergestellt) aus: made **out of** paper etc.
owing to	infolge, wegen dank: to be owing to = zurückzuführen sein auf, zuzuschreiben sein
thanks to	dank
with regard (respect, reference) to	hinsichtlich, bezüglich, was . . . betrifft

Exercises

A   Insert suitable prepositions:

1) The mechanical equivalent of heat is denoted .... the letter J.

2) She forgot to switch .... the current.

3) Acoustics deals .... sound.

4) The inertia of a moving train prevents it .... stopping as soon as the brakes are applied.

5) The radius of a circle bisects a chord .... right angles.

6) Form sentences .... the following words.

7) Change the following sentences .... questions and answer them.

8) A table is usually made .... wood.

9) Paper is made .... wood.

10) .... obtaining particularly bad results it is natural for a student to attribute the error to some purely mechanical defect.

B   When do you use . . .?

1) between: . . . .

2) among: . . . .

to 1) The prize for good speech was divided **between** Paul and Sheila.

to 2) The thunder echoed **among** the crags.

The money was distributed .... the crew. They knew that there was a private

eye .... the tourists. The cash was divided .... the two partners.

Switzerland lies .... Austria, Italy, France, and Germany.

High Priority:

Supply of Drinking and General Purpose Water

Supplying private households and industry with water is a task of national importance. This is, of course, recognized all the more in countries situated in arid zones, where the availability of water is a factor central to growth. It also applies, however, to the Federal Republic of Germany, although this country is in a generally favourable position with regard to the quantity of water available.

arid zones

In West Germany today, water requirements amount to more than 30 thousand million cubic metres per year. It is predicted that this figure will rise to 44 thousand million cubic metres by the year 2000. The share of water consumed by households in our country amounts to an average 250 litres per unit per day or 90 cubic metres per year (September 1980). But even in our country, the supply of water has become a problem from the point of view of the environment and resources, because the growing water requirements of private and industrial consumers are confronted on the one hand by partially diminishing stocks, indicated here and there by a gradual sinking of the ground water level. On the other hand, problems are posed increasingly by the quality of surface water.

diminishing stocks

As a result of increasing water pollution, and to some extent owing to the lack of terrain for carrying out the "soil passage" treatment, whereby biological activities take place in slow sand filters, new technologies using new preparation processes had to be found. Their common characteristic is the use of ozone before or after soil passage and systematic utilization of the natural biological decomposition processes, either in the subsoil or in synthetic sand filters. This method also provides a solution to the problem of toxic organic chlorine compounds.

Vocabulary

arid	dürr, unfruchtbar, trocken	with regard to	hinsichtlich
availability	Verfügbarkeit, Vorhandensein, Vorrat	requirement	(An)Forderung, Bedingung, Voraussetzung
brook	Bach	seep (through)	durchsickern
diminish	(sich) vermindern, abnehmen	share	(An)Teil, Beitrag, Beteiligung
encroachment	Über-, Eingriff; Beeinträchtigung	stock	Vorrat, Lager
interfere	sich einmischen; eingreifen; stören	toxic	giftig, Gift . . .
		treatment	Behandlung; Handhabung; Bearbeitung
lack (of)	Mangel (an), Fehlen (von)	utilization	Nutzbarmachung, Verwertung, Verwendung, (Aus)Nutzung
partial(ly)	teilweise		
predict	vorhersagen, voraussagen		

Questions

1) What are these 'arid zones' and where are they situated?
2) Why is there a growing concern about the water supply?
3) What is being done to filter out pollutants?

Paraphrase:

a) environment   b) diminishing stocks   c) lack
d) biological decomposition

Grammar

Indefinite Pronouns

Indefinite pronouns refer vaguely to persons and things not named.

Examples:

all	anything	everyone	neither	other
another	both	everything	nobody	several
any	each	few	none	some
anybody	either	many	no one	somebody
anyone	everybody	much	one	someone

....

When do you use . . . ?

1) every: ....

2) each: ....

3) some: ....

4) any: ....

to 1) **Every** child loves toys. (Each child . . .)
to 2) The headmaster talked to **each of** his teachers.
       We got 50p **each.**
to 3) She invited **some** relatives to her birthday.
to 4) We haven't **any** friends in Stuttgart.
       Is there **any** money left?
       If you have **any** complaints, please let me know.

Now form sentences with the other indefinite pronouns and explain their meaning.

all: ....

another: ....

anybody: ....

anyone: ....

anything: ....

both: ....

either: ....

everybody: ....

everyone: ....

everything: ....

few: ....

a few: ....

little: ....

a little: ....

many: ....

much: ....

neither: ....

nobody: ....

none: ....

no one: ....

one: ....

other: ....

several: ....

somebody: ....

Exercises

A   Insert 'some' or 'any':

1) Today, we can travel to .... country in the world in a very short time.

2) Is there .... thing that can exist without space, time, or matter?

3) Are there .... further questions?

4) Infinity is greater than .... number you can think of.

5) I have met this problem .... where before.

6) If you have .... thing to declare you must show it to the customs officer.

7) He wants to know .... thing more about it.

8) Time is .... thing that cannot be defined, but there is not ....
   thing that can exist without it.

9) They have .... knowledge of chemistry.

10) If he had .... notion about it, he would tell you.

B   Water Cycle

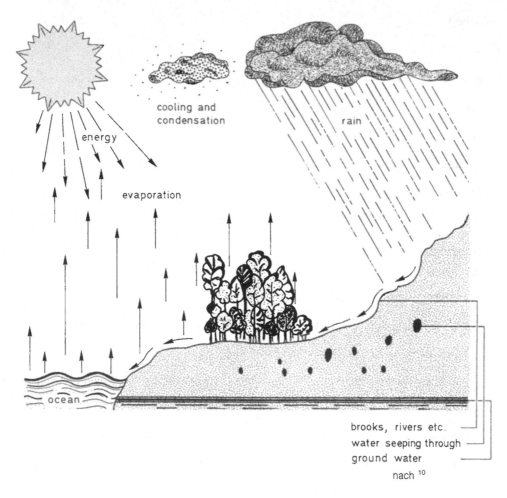

cooling and
condensation

energy

rain

evaporation

ocean

brooks, rivers etc.
water seeping through
ground water

nach [10]

Describe this cycle and point out where man has interfered with these natural processes and what could be done against further encroachments.

The Threat to Water

Water supplies are liable to contain a number of harmful pollutants, examples being bacteria, nitrates, phosphates, and detergents.

Bacteria may arise from human and animal excreta and, if they are not controlled, they can cause typhoid, paratyphoid, dysentry, and cholera. Bacteriological pollution is reduced by filtering and storing the water and by adding about one part per million of chlorine to it.

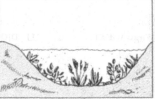

Water which has drained off from agricultural land may well contain nitrates and phosphates (output of private households: 70%, of industry 30%) since they are major constituents of fertilisers. These substances are a problem because they cause increased growth of weeds and algae in waterways and rivers. This in turn reduces the amount of dissolved oxygen in the water, and so fish and plant life may be adversely affected. It is also thought that nitrates may cause stomach cancer and increase the risk of a blood disease in very young children. Water may be treated by biological or chemical means to eliminate nitrates but these processes are expensive, and so the contaminated water is often simply diluted with water having a low nitrate content.

Some years ago, foaming was a familiar sight in rivers, lakes, and waterways. It was discovered that the cause of the trouble was the inability of bacteria to destroy detergents in sewage waters. This pollution has now been largely overcome by the use of 'soft' or biodegradable detergents which can be broken down, by bacteria, into simple molecules such as carbon dioxide and water.

Vocabulary

adverse(ly)	entgegenwirkend; ungünstig	detergent	Reinigungs-, Waschmittel; reinigend
alga	Alge	drain off	abfließen; versickern
degrade	abbauen	dysentry	Ruhr
constituent	(Bestand)Teil	excreta	Auswurf, Kot
contaminate	verunreinigen, vergiften, -seuchen	liable	verantwortlich für; neigend(zu)
content	Inhalt	sewage	Abwasser
		weed	Unkraut

Questions

1) What are the main pollutants and how do they get into the water?
2) What can be done to purify the water?
3) What do you know about purification plants? Outline the three main processes of purification: the physical, chemical, and biological method.
4) What are the effects of too much phosphate in the water?

Paraphrase:

a) detergent
d) foam

b) drain off
e) sewage water

c) weed
f) biodegradable

Grammar

The Conditional

The forms of the Conditional are used

1) after conditional clauses

If Paul had time, he **would go** to the course in glass blowing.
If Paul had time, he **would have gone** to the course in glass blowing.

2) ....

Jim said he **would carry out** the experiment.
Paula thought that she **would have done** everything.

Conditional Clauses
can be introduced by

unless (= if . . . not)
in case
on condition that
provided (that)

if (only)
even if, even though

wenn nicht; falls nicht, außer wenn
falls; für den Fall, daß
unter der Bedingung, daß
vorausgesetzt daß; unter der Voraussetzung, daß

wenn (nur); falls
selbst wenn

If-clause	**Main clause**
.... tense	.... tense
If he **learns** regularly,	he **will become** a good chemist.
.... tense	.... tense
If he **learned** regularly,	he **would become** a good chemist.
.... tense	.... tense
If he **had learned** regularly,	he **would have become** a good chemist.

Note: If I were rich, **I would travel** round the world.

Exercises

A Complete the following if-clauses:

1) If electricity (not to be) discovered, the telephone would not have been invented.
2) If I (to be) you, I should buy this chemistry book.
3) If I (to have) this substance, I could carry out the experiment.
4) If two quantities (to be) equal to a third quantity, they are also equal to each other.
5) If the children (to keep) quiet, they would have been permitted to see the movie.
6) If Newton (not to discover) the fundamental laws of mechanics, somebody else would probably have discovered them.
7) If $\sqrt{2}$ (to be) a rational number, we could express it as a fraction of two integers (ganze Zahl).
8) The law of the conservation of energy would be proved wrong if we (can) produce a perpetuum mobile.

B Translate and discuss

Gift im Rhein

Chemieunfall am Rhein

LUDWIGSHAFEN (lrs.) Etwa zwei Tonnen eines flüssigen Lösungsmittels sind nachts aus dem Werk Ludwigshafen der BASF durch das Kühlwassersystem in den Rhein gelaufen. Wie das Unternehmen mitteilte, handelte es sich um die Chemikalie Methyl-Pyrrolidon. Das Reinigungsmittel sei nur gering wassergefährdend und rasch abbaubar. Für Lebewesen im Rheinwasser oder für die Trinkwasserversorgung besteht nach Darstellung eines Unternehmenssprechers keine Gefahr. Den zuständigen Behörden sei der Vorfall gemeldet worden. Das Lösungsmittel ist laut BASF aus einem wegen technischer Arbeiten stillgelegten Betrieb zur Herstellung von Grundchemikalien in das Kanalsystem für Kühlwasser und dann in schwacher Konzentration in den Rhein gelaufen.

aus Stuttgarter Zeitung vom 31. 8. 87

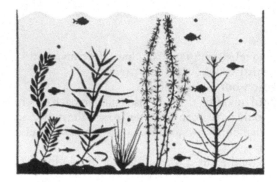

Surface Water Pollution

C   Describe these pictures and fill in the blanks.

grade A: much oxygen,
         few bacteria,
         wide variety of
         plants/fish

grade B:. . . . .

grade C:. . . . .

grade D: . . . .

aus [10]

Deep Concern: Ground Water

At the very top of the environmental scientists' list of concerns about pollution damage is something that most people probably believe to be safely beyond the reach of contamination:

ground water

This is water that lies buried from a few feet to *half a mile* or more beneath the land's surface in stretches of permeable rock, sand and gravel known as aquifers. In the U. S. there is five times as much water in such subterranean reservoirs as flows through all its surface lakes, streams and rivers in a year. While most ground water is believed to remain pure, concern is rising because it is one of nature's greatest non-renewable resources. Unlike surface water or the air, ground water is all but impossible to purify once it has become chemically polluted.

Ground water is not exposed to the natural purification systems that recycle and cleanse surface water; there is no sunlight, for example, to evaporate it and thereby remove salts and other minerals and chemicals. Nor can ground water be counted upon to clean itself as it moves through the earth, for it scarcely "flows" at all. It can take a human lifetime just to traverse a mile. Once it becomes polluted, the contamination can last for decades.

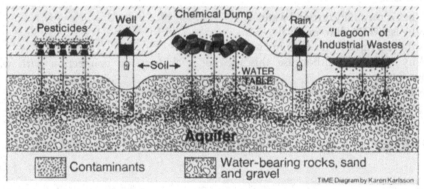

aus [8]

In the past, ground water was kept pure because the soil at the earth's surface could be counted on to act as a filtration system, a kind of geological "kidney" that would scrub out bacteria and other insoluble contaminants placed on or in the ground-before they could seep down to the water table, the ground water's upper limit. But this filtration system does not reliably screen out the waste chemicals that now leach into the soil from a variety of sources, including cropland that has been sprayed with pesticides, and industrial dumps like the pools into which liquid chemicals are placed so that the water they contain will evaporate.

A spokesman of the EPA (the federal Environmental Protection Agency in the US):

"We are not even sure if, not to mention how, chemical contaminants can be removed. It takes sophisticated testing just to determine if there are chemicals present. We cannot even begin to say how much of our drinking water, actual or potential, may have been contaminated. We will have to do a lot of detective work."

Vocabulary

aquifer	grundwasserführende Schicht	kidney	Niere
		leach into	ein-, durchsickern
beyond	jenseits	permeable	durchlässig
bury	begraben	screen out	herausfiltern
cleanse	reinigen	scrub	schrubben, reinigen
concern	Anliegen, Sorge	sophisticated	hochentwickelt, verfeinert
count on	zählen, rechnen mit		
cropland	Ackerbauland	subterranean	unterirdisch
dump	Müllabladeplatz		

Questions

1) What is meant by 'ground water'?
2) Why is it difficult to purify?
3) Is there a rapid flow of ground water?
4) Why does the author use this idea of a 'kidney' in connection with water purification?
5) Where does all this pollution come from and what is being done about it?

Explain in your own words:

a) permeable   b) traverse   c) cropland

Grammar

Gerund

It is by **measuring** the amount of ionization that radiation is most usually detected or identified.

a) **Measuring** the speed of this machine   – gerund as subject of the sentence
   required special devices.

   Die Messung der Drehzahl dieser Maschine erforderte Spezialgeräte.

   Um die .... zu messen, waren .... nötig.

b) The accumulator is charged **by passing** a current through it.

   – ....

....

....

c) They **succeeded in solving** the problem.

   – ....

....

d) An electric current is **capable of decomposing** water

   – ....

....

e) The mechanical rectifiers all have the **disadvantage of having** moving parts.

   – ....

....

f) They **stopped working** two hours ago.

   – ....

....

g) **It is no use trying** to convince him.

   – ....

....

....

h) We know **of him working** at a research institute.

   – ....

....

Try to gradually complete the following lists:

....

to insist on   afraid of   reason for
to prevent from   interested in   danger of
to object to   tired of   difficulty in

....

....

....

to love	it is useless	before
to continue	it is worth while	on
to avoid	to be busy	without

....

....

Exercises

A   Use gerunds:

1) After he had set up the circuit, he sent a current through it.
2) We use wire when we make coils.
3) We are sorry that we are lagging behind the other departments.
4) The scientist complains that the room is very hot.

Connect the following sentences:

a) We can change alternating current into direct current (use a rectifying tube).
b) We can change alternating current into direct current by using a rectifying tube.

5) These problems can be overcome (improve the design).
6) The mistake can be found (go through all the steps of the procedure).
7) The capacity of this cell can be increased (connect a number of negative plates and a number of positive plates in parallel).
8) It is possible to take perfectly stable nuclei and make them unstable (shoot into them various particles with the proper amount of energy).

B   Translate into German:

1) We know of your having achieved good results.
2) A great number of tests were carried out without any unforeseen breakdowns occurring.
3) Our students have already finished studying the initial data.
4) Nowadays engineers prefer using semiconductors for generating heat and cold.
5) The engineers intended applying a new method of welding.

C   Translate into English:

1) Fremdsprachen effektiv zu lernen erfordert viel Geduld.
2) Durch Erhitzen einer Substanz können wir ihre Temperatur bis auf den Siedepunkt erhöhen.
3) Beim Arbeiten mit radioaktiven Isotopen ist Vorsicht geboten.
4) Dieser Motor benötigt alle drei Monate, der neueste alle fünf Monate eine Durchsicht.
5) Ich erinnere mich daran, wie er im Regen stand.

The Silent Scourge

Nowadays the devastation brough by rains is silent, invisible, and pervasive. The killer is called acid rain. It is a particularly modern, post-industrial form of ruination, a blight as widespread and careless of its victims, and of international boundaries, as the winds that disperse it. Says Canadian Minister of the Environment John Roberts: "Acid rain is one of the most devastating forms of pollution imaginable, an insidious malaria of the biosphere."

Time[8]

The consequences are, indeed, numbingly real.

In the northeastern US, Canada, and Northern Europe, acid rain is reducing lakes, rivers, and ponds to eerily crystalline, lifeless bodies of water, killing off everything from indigenous fish stocks to microscopic vegetation.

In the process, acid rain is suspected of spiriting away mineral nutrients from the poor soil upon which forests thrive. Acid rain's corrosive assault on buildings and water systems costs untold millions of dollars annually. In addition, it is believed to pose a substantial threat to human health.

This assessment is not universally shared. The main dissenters are, predictably, cost-conscious government officials and lobbyists for the utility industry, whose coal- and oil-burning power plants are widely believed to be predominant contributors to the still obscure atmospheric chemistry that produces acid rain.

Indeed, precisely how acid rain forms in the atmosphere is still a mystery to scientists. Such natural processes as volcanic eruptions, forest fires, and the bacterial decomposition of organic matter can produce the acidic sulfur and nitrogen compounds that form acid rain. But most experts believe that the current problem is traceable to electrical generating plants, industrial boilers, and smelting plants that release sulfur dioxide (SO_2) and nitrogen oxides (NO_x) into the atmosphere, as well as acidic soots and traces of toxic metals such as mercury and cadmium. When they are vented into the air by tall smokestacks, molecules of SO_2 and NO_x are caught up in prevailing winds, where they interact, in the presence of sunlight, with vapor to form dilute solutions of nitric and sulfuric acids – or acid rain.

The worst acid rainfall measured in the US, pH 2.3, had about 1000 times the acidity of pure rain, pH 5.6.

7	6	5	4	3	2	1	0
NEUTRAL 7.0-Distilled water	6.6-Milk	5.6-Clean rain (any lower pH is acid rain)	4.5-Most fish die	3.0-Apples	2.2-Vinegar 2.3-Record rain in Kane, Pa.	1.2-Stomach acid	**ACIDIC**
			ACIDITY ON THE pH SCALE				

The pH scale* is, as you know, scientific shorthand for measuring acidity and alkalinity.

A neutral solution is pH 7. Because pH values are logarithmic, pH 1 is ten times as acidic as pH 2, pH 2 100 times as acidic as pH 3, and so on.

---

* verändert nach TIME, November 8, 1982, No. 45, S. 38f.

Vocabulary

assault	Angriff	prevailing	vorherrschend, vorwiegend
assessment	Bewertung, Einschätzung	scourge	Geißel, Plage
blight	Gifthauch, schädlicher Einfluß	shorthand	Kurzschrift
		smelt	schmelzen
boundary	Grenze	smokestack	Schornstein
devastation	Verwüstung	soot	Ruß
disperse	verteilen; ver-, ausbreiten	spirit away	wegzaubern
		suspected	verdächtig(t)
dissenter	Andersdenkende(r)	threat	Bedrohung
eery,ie	unheimlich, furchterregend	thrive	gedeihen, blühen
		traceable	nachweisbar, auffindbar
indigenous	einheimisch; angeboren	utility	Nutzen, Nützlichkeit; nützliche Einrichtung; Gebrauchs. . . Versorgungs . . .
insidious	heimtückisch; tückisch		
numb	erstarren(lassen), betäuben; starr, taub, betäubt	vent	(Abzugs)Öffnung, Auslaß
nutrient	Nährstoff	vent	abgeben
obscure	dunkel, unklar		
pervasive	durchdringend; überall vorhanden		

Questions

1) What are the signs of acid rain in places such as forests, towns, etc.?
2) What can be done to combat the bad effects of acid rain?
3) What are the general beliefs as to the cause of acid rain?
4) 'Let's fight the sources and not the symptoms'. Explain this statement.

Elucidate:

a) pervasive   b) insidious malaria   c) corrosive assault
d) mystery   e) volcanic eruption   f) current problem

Another Heritage at Risk

Read the text and discuss its content:

Architects, builders and sculptors in Europe were once content to protect their buildings and monuments from the ravages of wind and rain by rubbing them with linseed oil and various waxes. That was 300 years ago, long before the atmosphere over European capitals became foul with the corrosive by-products of industrialization. Nowadays there are few public works of art or architecture that could be saved by such simple methods.

In London, virtually every major historic building is suffering the ill effects of acid rain, from Westminster Abbey to the Tower of London. At St. Paul's Cathedral, started by Christopher Wren in 1675, up to an inch of Portland limestone has been worn away in some places, mostly by sulfur dioxide and other chemicals carried by London's famous rain and fog. The northwest tower of the façade of the cathedral was recently cleaned and renovated at a cost of $ 590,000. In the Netherlands, statues and other decorations on the outside of the 457-year-old 'S Hertogenbosch Saint John's Cathedral are "melting away like lollipops," according to Herman Teering, an architect and restoration expert. Dry particles of sulfur dioxide, cautions Jan Feenstra, a Dutch acid rain expert, can even infiltrate simple ventilation systems, playing havoc with old paper and books in museum archives. Sulfur dioxide in the air also creates black encrustations, called "stone cancer," that have riddled Amsterdam's Royal Palace on Dam

Damaged statue of Cologne Cathedral

Square, as well as the nearby Rijksmuseum. All told, acid rain-related destruction to historic Dutch monuments and buildings totals an estimated $ 10 million each year. In West Germany, damage to the Cathedral of Cologne alone costs $ 2 million a year to restore. In Rome, locally generated atmospheric pollution from auto exhaust fumes is blamed for defacing the marble relief of Trajan's column. So too with the deterioration of a 15th century sculpture by Jacopo della Quercia at San Petronio in Bologna.

Stockholm's 13th century Riddarholm Church's distinctive cast-iron spire had suffered so much corrosion from acid rain that it had to be replaced in the 1960s. Tord Andersson, conservator at the Swedish Central Office of National Antiquities, has now turned his attention to the church's rapidly deteriorating sandstone Royal Burial Chapel, where, he says, "we think the main problem is acid rain." Why? "Because the deterioration accelerated in the 20th century." Adds Andersson: "Acid rain is seldom the total problem, but if you can deal with it, you can minimize other troubles."

A team of experts from University College in London has been carefully monitoring the amounts of sulfur dioxide and other pollutants in local rainfall. The preservation specialists, including Robert Potter, Surveyor to the Fabric at St. Paul's, are also experimenting with a variety of preventive treatments. The simplest involves washing the stone with clean water sprayed from nozzles positioned at short intervals in a small-diameter pipe. More sophisticated is a new stone preservative called Brethane. When sprayed on stonework, it leaves a thin, colorless film that is absorbed and forms a protective coating. The treatment, however, is expensive ($ 170 per sq. yd.) and irreversible. Stockholm's Andersson is working with a consolidant called Silica-Ester, a compound similar in makeup to natural cement (lime) in the stone, that retards internal decay.

Such solutions are, at best, stopgap measures. "I don't think we can count on correction of the damage," says Andersson. "We must change the content of air pollution." Indeed, the deterioration of some 100 medieval limestone churches on the Swedish island of Gotland in the Baltic became so acute that antiquities officials successfully forced local utilities to burn only low-sulfur fuel oils, a reform that has become law throughout Sweden. "You can reduce the problem with conservation," warns Andersson, "but you can't stop it." The patrons of Europe's rich, ever diminishing cultural and religious heritage would do well to take note.

Time, November 8, 1982, No. 45

Vocabulary

blame for	verantwortlich machen für	heritage	Erbe
cast-iron	Gußeisen	irreversible	nicht umkehrbar
caution	(ver)warnen	linseed	Leinsamen
decay	Verfall, Zerfall; Fäulnis	marble	Marmor
deface	entstellen, beschädigen	nozzle	Mundstück, Mündung, Düse
deterioration	Verschlechterung	ravage	Verwüstung, Verheerung
distinctive	unterscheidend; kennzeichnend	retard	verzögern
		riddle	durchlöchern
encrustation	Überkrustung, -ziehung; Bedeckung	sculptor	Bildhauer
		spire	Spitze(Kirchturm-)
		stopgap	Lückenbüßer, Notbehelf
exhaust	Abgas; Auspuff		
havoc	Verwüstung, Zerstörung		

Grammar

Participle

a) This is the road **leading** to the lake.   – instead of a relative clause

Dies ist die Straße, die zum See führt.

b) **Wishing** to find the solution, she repeated the experiment.   – ....

....

c) **While reading,** he fell asleep.   – ....

....

d) **Though being** frustrated, they made another effort.   – ....

....

e) The neutrons split the uranium atoms, **(thus) releasing** enormous quantities of heat.   ....

....

f) I **saw** the instrument **getting** hot.   – ....

....

g) The assistant **sat** there **reading** the   – ....
instructions

....

h) They **kept** me **waiting.**   – ....

....

To f: **To g:**   **To h:**
to see, watch, notice,   to sit, remain, go,   to leave, keep, etc.
feel, hear, etc.   stand, come, etc.

....

Now form participle constructions, using the verbs given above:

....

....

....

Translate and discuss the following participle constructions:

1) This part of the country is called East   – ....
Anglia, **Suffolk being the southern
part** and Norfolk the northern part.

....

He was sitting at the table, **his arms
resting upon his open book.**

....

2) We walked about a little, **he showing   – ....
me** first the church and then the
school.

....

3) **There being** nothing else to do,   – ....
we went home.

....

4) We'll go, **weather permitting.**   –   ....

....

5) The wood is very beautiful **with the**   –   ....
   sun shining through the trees.

....

Vocabulary

a) generally speaking   allgemein gesprochen

   strictly
   properly speaking   genau genommen

....

b) concerning . . .   betreffs . . ., hinsichtlich . . .
   regarding . . .
   including . . .   einschließlich
   considering . . .   in Anbetracht . . ., angesichts . . .
   excepting . . .   ausgenommen . . .

....

c) considering (that) . . .   in Anbetracht dessen, daß . . .
   granting (that) . . .   zugegeben, daß . . .
   supposing . . .   angenommen, (daß) . . .
   providing . . .   vorausgesetzt, daß . . .

....

Now translate:

Jeder, mich selber nicht ausgenommen, muß sich an den Kosten beteiligen.

....

Angenommen, ich ginge wirklich, was würden Sie sagen?

....

Exercises

A   Use participle constructions:

1) As rubber is a good insulator, it is often used in cables.
2) Radioactivity was discovered; we made great progress in atomic physics.
3) An electric motor is a machine which converts electrical energy into mechanical
   energy.
4) If we speak strictly, there exist no such bodies in nature.
5) Mercury is used in barometers, because it has a great specific gravity.

6) The physical constant that characterizes the material of the conductor is called specific resistance.
7) Since he saw that the spark plugs were dirty, he cleaned them.
8) The earliest transistors were rather unreliable and could only be employed in very simple circuits.
9) The instrument partly lost its former efficiency as it had been used for a long time.
10) He used time to the best advantage; he completed his engineering course in four years.

B Translate into English, using participle constructions:

1) Wenn das Ergebnis des Experiments annehmbar ist, muß es veröffentlicht werden.
2) Da die Körper mehr Energie abstrahlen, als sie aus der Umgebung aufnehmen, kühlen sie ab.
3) Ich sah ihn, wie er den Bericht schrieb.
4) Obwohl er allein war, war er immer zufrieden.
5) Nachdem sie das Experiment gemacht hatte, verließ sie das Labor.
6) Ich saß in meinem Arbeitszimmer und schrieb einen Brief.
7) Die Buben, die den Hügel herabkamen, schlossen sich ihren Kameraden an.
8) Als das Abendbrot fertig war, gingen wir hinein.
9) Als die ganze Arbeit getan war, verließen wir das Institut.
10) Er hat sehr damit zu tun, alles in Gang zu halten.
11) Wer hat das Wasser laufen lassen?
12) Wir fühlten, wie der Boden erbebte.

C Translate into German:

1) This defect having been removed, the machine operated satisfactorily.
2) Following the instructions, you cannot make a mistake.
3) Most of the girls ran away leaving the others behind.
4) There being little evidence, the hypothesis couldn't be neither confirmed nor refuted.
5) We sat reading.

D Translate into English:

Das Schwefeldioxid spielt vermutlich eine wesentliche Rolle bei der Schädigung des Waldes. Es ist jedoch nicht mehr so einfach, den Zusammenhang zwischen dem Waldsterben und den industriellen Abgasen nach-zuweisen.	**probably**
Früher konnte man Schäden an Bäumen, etwa in der Umgebung einer Erz-Hütte, direkt erkennen.	**verify**
Heute treten Krankheiten in weit entfernten Gebieten auf, da die Schadstoffe durch hohe Schornsteine und den Wind weithin verteilt werden.	**smelting works** **smokestack**

Have a look at these controversial statements given by experts. Translate them and assess their implications:

„Fossile Energie durch
Kernenergie ablösen"

„Wir werden die Umwelt-
verschmutzung so niemals
in den Griff bekommen"

„pH-Wert im Boden
ist sehr wohl gesunken"

„Die Niederschläge sind gar
nicht saurer geworden"

„Es gibt einen
unbekannten Faktor
bei den Waldschäden

„Es genügt nicht, den
sauren Regen zu stoppen"

„Die SO$_2$-Leute tragen
nur einen Glaubens-
satz vor sich her"

E   Acid Rain

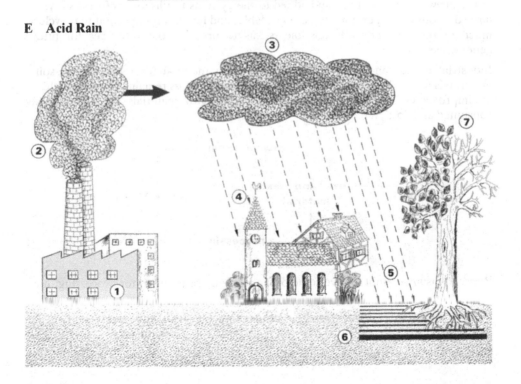

Translate the the following sentences:

1) Der Schwefel in Kohle und Öl verbrennt zu Schwefeldioxid.
2) Das Schwefeldioxid entweicht
3) und verbindet sich mit Sauerstoff und Regenwasser zu Schwefelsäure.
4) Die Säure greift Gebäude an
5) und dringt in den Boden ein.
6) Die Übersäuerung (hyperacidity) zerstört das biologische Gleichgewicht im Boden.
7) Pflanzen sterben ab.

Recycling

Recycling is essentially re-using, and it includes the reclamation of waste material and its reprocessing. Recycling is not a new idea. The 'rag-and-bone merchant' used to be a familiar figure, there have been scrap-metal dealers for many years, and waste vegetable and animal materials have long been composted by gardeners and farmers to return important chemicals to the soil.

Perfect recycling occurs in nature, e. g. in the carbon, nitrogen, and water cycles. In natural cycles like these the material passes endlessly through the cycle. We have 'interfered' with the nitrogen cycle but fortunately there is a large reserve of nitrogen in the air which can be fixed and added to the cycle as fertilisers. Unfortunately, no large deposits of copper, tin, etc., are available, and hence it is important to recycle as much of these materials as possible, so as to preserve our resources for future generations.

Industrial cycles can never return **all** the waste material for reprocessing; some material is always lost from the cycle so that an overall loss remains. However, this is an improvement on the 'open' system in which raw materials are continuously consumed and 'lost', i. e.

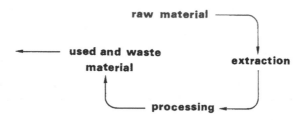

Recycling is an attempt to save some of the waste by returning it for processing, i. e.

Note that material which has been recycled does not need to go through the extraction stage. The extraction stage often consumes large amounts of energy, e. g. electricity, and so recycling also saves energy in addition to saving raw materials. It is less expensive to recycle steel from tin cans than to produce it from imported iron ore. Similarly, the production of copper from ore may take up to ten times the quantity of electricity needed to produce the same quantity of the metal from scrap. Aluminium requires 15 000 kilowatt-hours of electricity per ton from bauxite (the ore) but only 3000 kilowatt-hours per ton of scrap.

In future, the recycling of both industrial and domestic waste will become increasingly important as the technology for separating and treating such waste improves, and the processes become more economic (i. e. the cost of recycled material compares favourably with that of obtaining the same substance from raw materials).

In order to achieve this aim, positive pressure, co-operation from industry (particularly the manufacturing industries), and a better attitude from the individual are all needed.

For example, products could be designed and manufactured so as to be more easily recycled. Consider a simple metal container, which might typically be made from a mixture of products such as a body of tinned steel with a soldered seam (solder contains lead and tin) and aluminium ends. Such a container is difficult to recycle because the different components have to be separated. If the same container were to be made entirely of aluminium, it would be easy to recycle it.

A summary of the main advantages of recycling waste:

1) A saving of resources: trees, metals, oil, energy.
2) Improvement of the environment: no unsightly dumps, less use of valuable land.
3) Saving money: smaller bills for taxpayers.

4) ....

The Department of Environment Informs

Recycling

From waste disposal to recycling –
Energy and raw material from refuse

New waste recycling techniques can lead to the more effective use of the energy and raw materials contained in refuse. They can help to solve key problems in industrialized countries:

1) More effective disposal of refuse with less pollution. Waste material is growing in amount and becoming increasingly problematic in its composition. Hence, these recycling techniques can help to protect the environment;

2) By making use of the raw material content they can help to conserve valuable and scarce raw materials; to an increasing extent wastes, in particular industrial wastes, contain important raw materials, especially precious metals, which can be recovered by means of new techniques. Therefore, waste must be classified as "raw materials in the wrong place";

3) By making use of the energy content of the refuse they can help to ensure the future energy supply; today the thermal value of waste in many big cities already approximates to that of raw lignite. When the waste arising in conurbations is completely incinerated and its thermal power is utilized, approximately 8% of the local power requirements in these areas can be met. If the heat itself is used, it is expected that roughly the same proportion of the heat requirements for room heating and hot water can be satisfied.

Vocabulary

approximately	annähernd, ungefähr	rag-and-bone merchant	Alteisenhändler, 'Lumpensammler'
conurbation	Ballungsraum, -gebiet	reclamation	Rückforderung; Neugewinnung
disposal	Verwendung; Beseitigung	refuse	Abfall, Müll
domestic	häuslich, Haus. . .; inländisch	reprocess	wiederverwerten, -verwenden, -verarbeiten
incinerate	einäschern, (zu Asche) verbrennen	scrap-metal	(Eisen) Schrott, Alteisen
incineration	Einäscherung, Verbrennung	seam	Naht
lignite	Braunkohle	solder	löten
precious metal	Edelmetall	tin	verzinnen; eindosen
		unsightly	unansehnlich, häßlich

Questions

1) What do we understand by the word 'recycling'?
2) What is characteristic of natural cycles?
3) Sum up the advantages of recycling.
4) What can be done to bring about a change in the attitudes of the individual and of industry regarding recycling?
5) What are its disadvantages?

Explain:

a) 'rag-and-bone merchant'   b) 'open' system   c) extraction stage
d) soldered seam

Contrast the classical refuse techniques of:

– **dumping, composting** and **incineration**
  with advanced waste utilization technologies such as
– **pyrolysis** and **raw material recovery.**

Grammar

Differences Between British and American English.

Different words for the same things:

British	American	German
aerial	antenna	....
aluminium	aluminum	....
autumn	fall	....
bonnet	hood	....
chemist's shop	drugstore	....
cupboard	closet	....
draught(sman)	draft(sman)	....
to ensure	to insure	....
to gauge	to gage	....
gramophone	phonograph	....
lift	elevator	....
lorry	truck	....
manoeuvre	maneuver	....
metre	meter	....
mould	mold	....
petrol	gasoline, gas	....
programme	program	....
railway	railroad	....
shop	store	....
solicitor	attorney	....
spanner	wrench	....
tap	faucet	....

tin	can	....
tram(way/car)	street-car	....
tyre	tire	....
underground (tube = colloq.)	subway	....
warehouse	department store	....
windscreen	windshield	....
wireless	radio	....

. . .

Find other examples and write them down.

The same word, but a different meaning:

	British E.	American E.
bureau	Schreibtisch	Kommode
corn	Getreide	Mais
schedule	Liste, Tabelle, Verzeichnis	Lehr-, Arbeits-, Stundenplan, Fahrplan

Different pronunciation

schedule	(Br.)
	(US)
bird	(Br.)
	(US)

. . .

Simplifications in spelling

Br.: travel – travelling, travelled, traveller
US.:   – traveling, traveled, . . .

Br.:	colour	honour	behaviour	
US.:	color	honor	behavior	....
Br.:	theatre	centre	metre	
US.:	theater	center	meter	....
Br.:	offence	defence		
US.:	offense	defense	....	

Simplifications of consonants

wagon (Br. waggon); catalog (Br. catalogue); tarif (Br. tariff); check (Br. cheque); plow (Br. plough); tho (Br. though); thru (Br. through);
. . .

Different expressions

Stick no bills! (Br.)
Post no bills! (US)

. . .

Differences in grammar

Explain the examples:

His brother is taller than **him**.

He has two sister-in-**laws**.

beautiful, beautiful**ler**, beautifull**est**;

He **hasn't** any money (Br.)

He **doesn't have** any money (US)

In common speech:

Have you done it **proper?**

We **don't** see **no**body.

a coupl**e b**oys   (Br. a couple **of** boys)

at a quarter **of** eight   (Br. at a quarter **to** eight)

Differences in stress

alloy – Legierung   (   ) (Br.)   (   ) (US)

. . .

Find more distinctive features of the 'two languages' and write them down.

. . .

Translate the following sentences and point out the different meaning of 'but':

A line has **but** one dimension – length.

There is **but** one way out.

She is **but** a student.

. . .   1) **but** = ...

He came **but** last week.

. . .   2) **but** = ...

He was **all but** drowned.

. . .   3) **all but** = ...

Exercise

Read, translate and discuss:

Recycling

Den Hausmüll neu verwerten

In einem Jahr produziert der Bundesbürger etwa 250 Kilogramm Müll. In den Städten fallen je Einwohner und Jahr zusätzlich 150 bis 250 Kilo Gewerbemüll an. Je Bürger kommen jährlich 75 bis 125 Kilo Klärschlamm hinzu. In den Städten sind je Einwohner jährlich außerdem 50 Kilo Straßenkehricht, Markt-, Garten- und Parkabfälle zu beseitigen. Die jährliche Abfallmenge je Stadtbewohner liegt zwischen 600 und 800 Kilogramm. Die darin enthaltenen Rohstoffe sind groß. An der Berliner Technischen Universität rechnet Professor Jäger, daß städtischer Hausmüll fast zur Hälfte aus wiederverwertbaren Stoffen besteht. Ihre Nutzung setze Umstellungen voraus, die von der aktiven Mitarbeit der Bürger, getrenntem Einsammeln bis zu neuen technischen Verfahrensweisen reichen.

(aus Sonntag Aktuell vom 23. 1. 83)

Pesticides, Fungicides, Herbicides

Chemists have to find answers to many problems facing modern civilization, especially increasing demands for energy and food. The world population continues to grow at an alarming rate, and it is becoming more and more difficult to provide anything like enough food. At least half of the people in the world are suffering from some kind of food shortage. This is particularly serious in the developing countries, where natural resources are often limited or unexploited and where technology is improving only slowly. Each year there are some 80 million extra mouths in the world to be fed.

One solution to the problem would be to use more land for agricultural purposes, so that more food crops could be grown. This is not a long-term answer, however, for most of the undeveloped land which remains is not suitable for agricultural use. The problem is that the number of mouths to be fed is increasing but soon no more land can be made available to grow food.

The problem can only be solved (or partially solved) by improving the amount of food which can be obtained from a given area of land and by taking steps to reduce the birth rate. Biologists are involved with these problems, e. g. in helping to control the birth rate and in developing new kinds of wheat, maize, etc. which produce better yields of food.

Chemists are concerned with the development of fertilizers, pesticides, herbicides, fungicides, and various minerals and vitamins which are added to the food given to farm animals.

Pesticides kill insects and other pests which attack food crops and reduce their yield.

Fungicides kill fungi which cause plant disease.

Herbicides kill weeds which would otherwise prevent some soil nutrients, water, air, and light reaching the crop.

Similarly, farm animals produce food more efficiently if they are given supplements in their diet, and are kept free from disease by the use of sprays, dips, and injections.

It goes without saying that steps have to be taken to ensure that agricultural chemicals are not wrongly used.

Some fertilizers, if used in excess and in the wrong situation, can be washed into water courses and lead to pollution.

Careless use of pesticides can also kill **helpful** insects and other animals (e. g. bees, ladybirds etc.) as well as harmful pests.

If a pesticide is **persistent** (i. e. is not broken down quickly to harmless substances by animal cells, plant cells, bacteria, or ultraviolet light from the sun) it may become concentrated in a food chain.

Chemists are constantly improving chemicals used in agriculture, and people are being trained more thoroughly in the correct use of these chemicals.

Fertilizers

Plants use carbon dioxide and water to form carbohydrates by the process known as photosynthesis. However, they also need a number of chemical elements as nutrients.

The major nutrients are

- nitrogen,
- phosphorus,
- potassium

and to a lesser extent calcium, magnesium, and sulphur.

Nitrogen stimulates the growth of leaves,
phosphorus aids stem and root growth, whilst
potassium helps the plant to make sugar.

Besides the major nutrients, plants need traces of other elements for their efficient growth. The so-called trace elements include boron, cobalt, copper, iron, manganese, molybdenum, and zinc, and most soils have adequate reserves of them.

Many farmers throughout the world simply rely on animal manure to keep their land fertile. However, manure has only limited amounts of the nutrient elements, and it has been found that the use of chemical fertilizers can greatly increase crop yields.

Thus, fertilizers play a very important part in the production of crops and animal fodder such as grass, etc. However, as with all resources, they should not be wasted by excessive use, and care is needed to limit the risk of pollution.

Vocabulary

boron	Bor	rate	Geschwindigkeit;
dip	Desinfektionsbad		Maß, Kurs; Betrag;
fodder	Futter		Grad
fungicide	pilztötendes Mittel	thorough(ly)	gründlich
herbicide	pflanzentötendes	trace element	Spurenelement
	Mittel	unexploited	nicht ausgebeutet,
ladybird	Marienkäfer		-abgebaut, -verwertet
long-term	langfristig	wheat	Weizen
maize	Mais		
manure	Dünger		
pest (insect-)	Plage, (Schädling)		
pesticide	Schädlingsbekämp-		
	fungsmittel		

Questions

1) What are pesticides, fungicides, and herbicides; what are they used for, and why?
2) What are their possible dangers?
3) How do fertilizers affect plants?
4) How important are they in agriculture?
5) What should be done to prevent misuse?

Elucidate:

a) alarming rate   b) food shortage   c) supplement   d) food chain
e) photosynthesis   f) throughout   g) excessive use

Grammar

Contact Clause

Compare the two following sentences and explain why a relative pronoun must be used in the one case and not in the other.

This is the man .... I saw yesterday.

....

....

This is the man **who** gave me the book.

– They held debates on anything .... they liked.
Sie veranstalteten Diskussionen über alles, was sie nur wollten.

– ....

– He is not the man .... he was before he came here.

– ....

....

– This copy is the only one .... there was in the shop.

– ....

....

– He did it in the way .... I would have done it myself.

– ....

....

Exercises

A   Form contact clauses from the following sentences:

1) The solution seems satisfactory; the student proposes it.
2) The material will be supplied regularly; further research depends upon it.
3) Evidently the method could be of great value; the author has outlined it.
4) There is one point; we have not dealt with it so far.
5) The four steps are continually repeated; the procedure consists of them.
6) The conclusions must be verified; we have arrived at them.
7) The device seems very effective; you have mentioned it.
8) The papers are available in our library; the author refers to them.

B   Translate into English:

Natur- und Mineraldünger sind kein Gegensatz

Mineraldünger enthalten genau die gleichen Nährstoffe, die von der Natur aus in einem gesunden Boden vorhanden sind und die von den Pflanzen gebraucht werden.

Das sind u. a. Stickstoff, Phosphat, Kali, Kalk und Magnesium. Aber auch Spuren-nährstoffe, wie Bor, Eisen, Mangan und Kupfer.

Stickstoff gewinnt man dank der Ammoniak-Synthese aus der Luft. Phosphat, Kali, Kalk und Magnesium werden bergmännisch dort abgebaut, wo sie in großen Mengen vor Jahr Millionen von der Natur abgelagert wurden. Mineraldünger sind also nichts Künstliches, sondern etwas ganz Natürliches.

C   Translate into English:

Keine ausreichende Ernährung ohne Mineraldünger

In der Vergangenheit wurden viele Versuche unternommen, den Nährstoffentzug des Bodens wieder auszugleichen: durch Fruchtwechsel (**crop rotation**) – in einem Jahr Getreide, im nächsten Jahr Rüben, dann wieder etwas anderes – oder durch die Brache (**fallow**), also das zeitweise 'Ruhenlassen' der Äcker. Alle Versuche blieben schließlich ohne den gewünschten Erfolg. Selbst die Verwendung von Stallmist konnte den Nährstoffentzug nicht ausgleichen. Die Erträge nahmen weiter ab, während die Bevölkerung sprunghaft wuchs. Neuere Untersuchungen wissenschaft-licher Institute zeigen: Auch die 'alternativen' Anbaumethoden produzieren oft nicht einmal die Hälfte der Erträge, die bei bedarfsgerechtem Einsatz von Mineraldüngern und Pflanzenschutzmitteln möglich sind.

D

Ohne chemischen Pflanzenschutz wäre unsere Ernährung eine Katastrophe.

Ägyptische Wanderheuschrecke (Schistocerca gregaria), die in riesigen Schwärmen verheerende Fraßschäden anrichtet

Discuss the implications of this picture and translate the following text:

Unsere Ernährung hängt fast ausschließlich von den Ernte-Erträgen (**yield**) ab.

Der Ackerbau hat an der Gesamt-Nahrungsmittelproduktion der Welt einen Anteil von 97 Prozent.

Auf die Fischerei entfallen (**to fall to**) dagegen nur zwei Prozent, auf die Weidewirtschaft sogar nur ein Prozent.

Niemand wird deshalb ernsthaft in Frage stellen, daß wir die Pflanzen, die unser täglich Brot sind, schützen müssen.

Noch vor 65 Jahren starben in Deutschland 700 000 Menschen an Hunger (**to starve**), weil die Ernte durch Schädlinge (**pests**) und Pilz-Krankheiten (**fungus disease**) fast vollständig vernichtet wurde.

Chemistry is when . . .

Every Picture Tells a Story

Have a go at this one!

aus [10]

Chemie ist, wenn kein Wurm drin ist.

Pilzkrankheiten und Schädlingsbefall verhindert die Chemie. – Chemie auf Ihrer Seite.

Es informiert Sie die Initiative „Geschützter leben" der Chemischen Industrie, Karlstr. 21, 6000 Frankfurt.

Chemie ist (auch), wenn . . .

. . . zum Beispiel am Bodensee Vögel tot vom Himmel fallen   Foto: J. Resch

Ganz Deutschland entsetzte sich am Fernsehschirm, als es im Frühjahr dieses Jahres sterbende Vögel regnete. Aus dem scheinbar so heiteren Himmel der Bodenseeregion zwischen Ludwigshafen und Lindau fielen innerhalb weniger Tage mehr als hundert Mäusebussarde und Sperber, Habichte und Falken, Eulen und Amseln, über vierzig Singdrosseln, dazu Lerchen, Blau- und Kohlmeisen tot zur Erde. Daneben verendeten Säugetiere wie Wiesel, Iltisse und Katzen.

Die Ursache des Massensterbens war schnell gefunden: Die Obstbauern des Landes hatten ihre Plantagen in diesem Jahr besonders intensiv mit endrinhaltigen Kampfmitteln gegen Wühlmäuse gespritzt. Tödliche Mengen davon fand man in den Lebern der obduzierten Vogelkadaver.

Die Behörden reagierten ungewohnt hart und prompt. Landwirtschaftsminister Gerhard Weiser erließ ein vorläufiges Endrinverbot für Baden-Württemberg, die Biologische Bundesanstalt in Braunschweig verbot im Juni den Vertrieb bundesweit. So dürfen die Sprühmittel „Mäuse-Kindrin 391", „Segetan Wühlmausmittel", „ST-M3" und „Wühlmausmittel Fagacid" nicht mehr in den Handel gebracht werden. Zudem wurden Schritte eingeleitet, auch die Anwendung der Mittel kurzfristig zu verbieten. Wer danach noch sprüht, kann nach Paragraph 25 des

Pflanzenschutzgesetzes mit einer Geld-
buße bis zu 10 000 Mark bestraft
werden.

Es wurde auch Zeit. Denn der Massen-
tod am Bodensee war keine Ausnahme.
Seit Jahren sterben Vögel nach Anga-
ben vom Bund für Umwelt- und Natur-
schutz (BUND) in Gebieten, in denen
Bauern endrinhaltige Mittel einsetzen;
so bereits 1959 nach einer Anti-Feld-
mauskampagne im Illertal und 1960 aus
gleichem Anlaß im Landkreis Biberach,
wo neben Eulen und Raubvögeln auch
Hasen und Rehe verendeten.

Die Dunkelziffer ist hoch, da Endrin
vorwiegend im Herbst gespritzt wird. Zu
dieser Zeit haben viele Vögel keine fes-
ten Reviere mehr. Sie sterben nicht am
Tatort, sind also als Endrinopfer nicht
mehr zu lokalisieren.

On the previous two pages are two pictures which have some connection with
chemistry.

Read the text; give an account of the information given and point out what different
effects chemistry can have.

What precautions must be taken?

Pollution: As Old As Mankind Itself

The problems we have to deal with today concerning conservation are not as new as
we would like to think. Man has apparently always suffered from some form or other
of pollution – although before now it was not as wide-spread and the damage was not
as destructive.

In earlier times people felt, for the most part, harmed by bad air: – the Greek
physician Hippocrates (460–377 B. C.) referred to the evil smells produced by tanning
and to the poisonous gases from silver smelting.

It appears, too, that Rome stank badly in certain parts; both Seneca (4 B. C. – 65
A. D.) and Pliny the Elder (23–79 A. D.) and later the physician Galenus (129–199
A. D.) complained of the bad air.

The Romans had the same problem with air pollution in their capital – although not to
the same extent – as we have in the 'Ruhrgebiet' today: the senators no doubt had
trouble with their white togas because the air was permanently full of soot.

King Edward I of England (1239–1307) had to move his seat of government from
London to Nottingham because his wife was allergic to charcoal smoke.

London air was particularly bad and in 1661 Samuel Pepys wrote in his diary that it
was a thick, filthy, stinking fog which could cause catarrh, coughing and consumption.

Early on in history people saw the resettlement of commercial enterprises as the only
way out of their plight.

In 1348 in Zwickau smiths were prohibited from using hard coal; from 1427 in Goslar
the owners of iron and steel works were no longer allowed to smelt ore anywhere near
the town. In Cologne in 1464 a lead works was closed down within two weeks, and in
Venice in the fourteenth century all works which released poisonous gases into the air
were not allowed to remain in the city's sovereign territory.

Through the centuries mankind has shown itself to be particularly unreasonable in its behaviour towards its environment; one need only think of the felling of enormous forest areas since the time of the ancients and the devastating results for both the climate and the environment.

Right up to the present day we have not learnt our lesson!

aus [12]

Comment on this picture and enlarge upon the following saying:

Des einen Dreck –
des andern Umwelt

Vocabulary

allergic	allergisch	evil	böse, übel
the ancients	die Alten (Griechen und Römer)	fell	fällen
		filthy	schmutzig, dreckig
charcoal	Holzkohle	hardcoal	Anthrazit, Steinkohle
conservation	Erhaltung, Bewahrung, Schutz	plight	(mißliche) Lage, Notlage
consumption	Verbrauch; Schwindsucht	resettlement	Wiederansiedlung, Neuordnung
enterprise	Unternehmen	tan	gerben, beizen

Questions

1) What sorts of problems did the Ancients have with pollution?
2) What were the measures the citizens took to counter these ill effects?
3) What comment does the author make on mankind? Do you agree? Give reasons for your answer.

Explain:

a) conservation
b) toga
c) allergic
d) diary
e) consumption
f) prohibit
g) sovereign territory
h) fell
i) devastate

Grammar

The Prop-Word 'One'   **Translate**

They had to do an oral exercise, not
a written one.   ....

From time to time the old rails must be
replaced by **new ones.**   ....

In these cases the prop-word 'one' is used to . . .

_____

He prefers a white wine to **a red one.**   ....

But:

He prefers white wine to **red.**   ....

Explain the difference.

_____

I don't like this dress.
Can you show me a better one?   ....

In this case the prop-word is used . . .

But:

We may distinguish several types.
The **simplest** and the **most frequent** is ....

_____

He was one of **the lucky ones.**   ....
(man, woman, person, people, etc.)

Explain.

Exercise

Translate into English:

1) Die Beleuchtung war die beste, die er je gesehen hatte.
2) Diese Ausgabe ist zu teuer; kann ich eine billigere haben?
3) Ist dies Zimmer das deines Bruders oder dein eigenes?
4) Ziehen Sie englische Apparate den amerikanischen vor?
5) Es gibt hierauf eine richtige Antwort und eine falsche.
6) Diese Grammatik ist die kürzeste, die ich kenne.

4 Chemistry – At Second Sight

Tracing Traces

Lieutenant **Stone** and Inspector Heller screech to a halt. Lights flashing and sirens blaring. A body lies in the gutter, a hit-and-run victim. A sergeant reports to Stone that the body was found fifteen minutes earlier. While they talk Heller goes over to the body and examines it. He notices something on the victim's trousers: **car paint.** He turns the corpse over and goes through the pockets; along with a wallet and some receipts he finds four small clear plastic bags containing a **white powder.** He calls to Lieutenant Stone and hands it over: "I think the car paint and the bags should be got to the lab right away."

Down the block **Kojak** is already on the scene. As he enters the seedy hotel he sees Stavros: "O. K. Stavros baby, what you got?" A young woman's body is sprawling on the carpet. Stavros tells him that **skin** has been found under the woman's fingernails and that **hairs**, a different colour from the woman's, are on the carpet. While forensic experts take samples of **finger-prints** and of the **soil** on her shoes, the Lieutenant looks around the room. Next to the bed he sees a **cigarette end**. Kojak: "Hey, you guys, get these skin and soil samples and hairs down to the lab. And you forensic men get your butts over here to take what I've found."

Steps go up two flights of stairs and Lieutenant **Columbo** stumbles into an upturned apartment, his coat flapping behind him. "I'm sorry I took so long getting here, Officer, but it was my wife, she . . . "Having made his apologies and excuses he picks his way through the broken furniture and as he does so he notices a **cheque** poking out from under a table and a broken pen not far from it. A glass lies tipped over and the **sweet smelling liquid** has soaked into the rug; but what really catches his attention is a **dark stain** near a chair. He bends down, dabs his finger on it and looks at it closely. Blood? "Er, excuse me, Officer, could you pick up that cheque and get it to the lab to find out whether it's a forgery, please? Oh, I almost forgot, there's a glass over here, check if poison is in it. Yeah, and that nasty looking stain. Take a couple of samples and have them analysed. Thanks, I'd appreciate it".

From these three cases we can see how even such great detectives as Kojak, Columbo and Co. are dependent upon the 'lab' to help them solve the crimes they are faced with.

The chemists in these labs are able to detect and identify even the merest trace of any substance handed over to them. With the help of mass spectrometry, electron microscopy, gas-chromatography, different kinds of spectroscopy, etc., they have actually become the most competent people to uncover crimes.

By working hand in hand with forensic experts and the police they have improved the quality and efficiency of police work and thereby increased the rate of crimes solved.

In front of you there is an unidentified substance. Let's see whether it's blood or not:

This can be ascertained by the 'Luminol* Reaction', the analysis which is most widely used. The following three separate storable solutions have proved the most efficient:

---

* (3-aminophthalhydrazide)

a) 8 g of NaOH in 500 ml of water;
b) 10 ml of a H_2O_2 solution (30%) in 490 ml of water;
c) 0.35 g of Luminol in 62.5 ml of 0.4 n NaOH, topped up with 500 ml solution.

If need be 10 ml of these three solutions are added to 70 ml of water and this mixture is used as a spray. This reaction must be observed in the dark, because a positive reaction, especially in cases of old traces of blood, is characterized by a light blue chemiluminescence.

However, it can be a disadvantage because the sprays have a relatively low rate of specificity. Chlorophyll, especially, (for example, the sap of trodden grass) shows a very positive reaction.

Vocabulary

ascertain	ermitteln, feststellen	rug	(kleiner) Teppich,
blare	heulen, brüllen		Vorleger
butt	(dickes) Ende, Hintern	screech	kreischen
		seedy	heruntergekommen,
chemilumi-nescence	Chemolumineszenz		schäbig
		sprawl	ausgestreckt daliegen
flap	flattern, (hin u. her bewegen, -schlagen), lose herunterhängen	stain	Fleck
		storable	zum Lagern geeignet, Lager. . .
forensic	gerichtlich, Gerichts. . .	tip over	umkippen
		trodden (to tread)	(zer)treten, (zer)trampelt
forgery	Fälschung	upturned	auf den Kopf gestellt,
gutter	Gosse, Rinnstein		umgeworfen, umge-kippt
hit-and-run	Fahrerflucht		
mere	bloß, nichts als, allein		
poke out	hervorschauen	wallet	Brieftasche
receipt	Beleg, Quittung, Rechnung		

Questions

1) What might have been the motive for the murder of the man with reference to the white powder found in his pocket?
2) What are the implications of the words in bold print in the 2nd case?
3) What type of man is Columbo: his personality, his methods of working, the way he deals with people? Compare him to Kojak and Stone.
4) Say in your own words why even these celebrities are so dependent on 'the lab'.

Paraphrase or give another word for:

a) screech to a halt
b) gutter
c) seedy
d) upturned
e) stain
f) forgery
g) the merest trace
h) competent people
i) efficiency

Grammar

The Infinitive

with "to":

a) **To measure** this current will be very   – used as subject
   difficult.
   Diesen Strom zu messen wird sehr
   schwierig sein.

b) The problem **to be discussed** in this   – used as ....
   paper is entirely new.
   Das Problem, das in diesem Artikel erörtert wird, . . .
   Das Problem, das in diesem Artikel erörtert werden soll, . . .
   Das in diesem Artikel zu erörternde Problem . . .

c) The failure caused **the designers to**   – ....
   modify the appliance.
   Der Fehler (Versagen) veranlaßte die Konstrukteure, das Gerät zu verbessern.
   Translate:
   Automatic control permits **this process to be speeded up.**

   ....

d) **The transformer** is known **to be**   – ....
   a useful apparatus.
   Man weiß, daß der Transformator ein nützliches Gerät ist.
   Es ist bekannt, daß . . .
   Bekanntlich ist der . . .
   Der T. ist **als** ein nützliches Gerät **bekannt.**
   Der T. ist **bekanntlich** . . .
   Translate:
   After this treatment, **the material is then said to be magnetic or magnetizable.**

   ....

e) There is a tendency **for these tiny**   – ....
   currents to cancel one another out.
   Es besteht die Tendenz, daß sich diese winzigen Ströme gegenseitig aufheben.
   Translate:
   For an isolated atom to alter its state, it must absorb a photon.
   The safety inspector must arrange **for safety devices to be installed** on all machines.

   ....

Exercises

A   Change the following sentences.

Example:
- It is very difficult to measure this current.
- To measure this current is very difficult.

1) It will certainly be possible for them to solve this problem.
2) It will hardly be possible to tell in advance what the results of this experiment will be.
3) It will be easy to determine the frequency of a radio wave.

B   Change the sentences by using the "for-construction" beginning with:

There is no need . . .; There is the tendency . . .; It will be necessary . . ..

1) Industry tends to be concentrated in a single region.
2) Cast iron tends to fracture under excessive tension.
3) They will have to study the fundamentals of electrical engineering.
4) It is not necessary that the operator knows the theory behind the operation of his machine.

C   Translate into German:

1) Electrons, atoms, and molecules are known to possess quantized magnetic dipole moments.
2) If a computer is required to make a calculation or to solve a problem, it has to be fed the necessary information.
3) This electronic computer is believed to be the most complicated one in our research laboratory.
4) The fundamental questions to be treated in the present paper can be broadly divided into two general categories.
5) This method permits information to be read by both man and machine.

D   Translate into English. Use infinitives.

1) Der zu zahlende Betrag ist ziemlich groß.
2) Hohe Temperaturen zu messen ist die Aufgabe von Thermoelementen.
3) Diese Vorrichtung gestattet es, die Spannung nach Belieben zu verändern.
4) Es ist wichtig, daß die Diskussion offen und ehrlich geführt wird.
5) Das Instrument, das von unseren Ingenieuren konstruiert werden soll, hat große praktische Bedeutung.

Chemistry in Sport

Olympic Games in Munich, 1972:

Once again the youth of the world meets for competitions characterized by speed, strength and stamina. It is, however, not only a competition of athletes, it is a competition of materials, too. The best example of this is the pole vault. Bob Seagran, who won the Gold Medal in Mexico in 1968, is in the newspapers because he is banned from jumping with a pole made of a newly developed chemical product. The reasons for this ban have been kept secret from most of the spectators in Munich. Today we know that he used a pole made of carbon fibre, a material unknown in 1972 and therefore not allowed.

Chemical research has also led to improvements in the construction of skis – especially of their racing tread. Today we smile if we watch films showing our grandfathers on skis made of solid wood. But even our modern skis contain a wooden core. It was in the 1950's that wood was gradually replaced by different materials. First, the tread was substituted by synthetics; second, the wooden core was completely covered with aluminium and eventually, chemical industry succeeded in producing a faster ski.

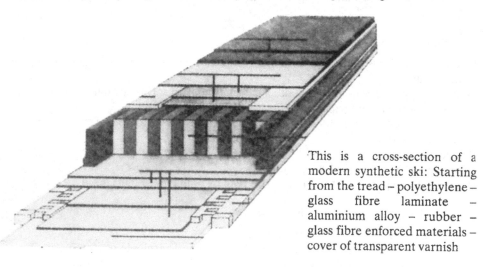

This is a cross-section of a modern synthetic ski: Starting from the tread – polyethylene – glass fibre laminate – aluminium alloy – rubber – glass fibre enforced materials – cover of transparent varnish

Glass fibre fabric was used for the construction of 60 fin dinghies for the Olympic sailing competition in Kiel. Since the greatest weight difference was less than 500 g, each participant had the same chance in terms of materials. The construction of rowing boats resembles far more the construction of sophisticated jet planes than the manufacturing of sports equipment.

Modern sports equipment is virtually a kaleidoscope of our plastic age. Take a sailing boat, for instance: The sails are made of synthetic fibres which keep their shape forever. The mast is no longer made of wood but of a specially treated material. Polyamid prevents the ropes from rotting and becoming mouldy. The hull is completely formed from synthetic resin and glass fibres. What remains to be discussed are the high costs for such products: some hundreds of D-Marks per kilo! No wonder then that a record-breaking bicycle made of carbon fibres and tested in the wind canal costs the princely sum of 25,000 DM.

aus [14]

Today's racing track is no longer made of yesterday's ashes. The combined efforts of chemists and engineers have succeeded in creating a synthetic surface, which doesn't become slippery even in the worst thunderstorms. Synthetic playgrounds and sports grounds show similar resistance. They can be used in both summer and winter, thus always allowing the athletes to train and improve their records.

Quite a different aspect concerning sports records is the athlete's diet. Each top athlete has to have his specific diet which often has to be supplemented with such things as vitamins, minerals and trace elements. Some sportsmen, however, try to increase their efficiency by doping – an illegal use of drugs. Hormone-like drugs (anabolica) improve the muscle growth, tranquilizers calm down hypersensitive

athletes whereas stimulants provide the necessary alertness. Doping has been prohibited for some years because it violates general rules of fairness. Here is another example of the chemist helping sports referees expose violations. At the same time, of course, he plays a vital role in keeping the sportsmen in good health.

Modern synthetics have improved sports and have led to new records and almost identical sports equipment – often at a price that has made the dreams of many a sportsman come true. Additionally they have also affected wide ranges of leisure sports. Compare the millions of skis sold each winter, the numerous tennis rackets, golf clubs, fishing rods, surf boards and the mountaineering equipment – neither of which could ever have been produced without the progress made in chemistry.

In many regions the skiing season lasts only a few months, which means empty rooms and lifts at a standstill during the rest of the year. But even here, chemistry has succeeded in helping the skiing centres with artificial snowflakes made of polyethylene. This artificial slope can withstand any weather conditions. Even skating rinks are made of synthetics – leaving them no longer dependent upon cold weather or energy saving because cooling is no longer necessary.

Yet chemistry also has its problems. For instance astroturf – an artificial lawn – which raises questions for athletes and spectators alike. Many environmentalists may be saddened by the idea of a lawn which stays green forever. Sportsmen get into trouble if the lawn is exposed to rain or fog; then the surface becomes slippery and players lose their balance – and self-confidence. Moreover, these synthetic materials affect the ankle joints in a different way from grass and wood, organic materials which man has got used to and adjusted to in all his movements in the course of millenia. Questions arise as to possible harmful effects on the human body. Therefore scientists analyse in a long series of examinations which pressures may have a harmful effect on joints, muscles and tendons when stopping and jumping on synthetic courts. One thing is certain: sport and sportsmen will be more and more dominated by synthetics, but the effects on sportsmen cannot be fully determined at the moment.

Vocabulary

alertness	Wachsamkeit; Flinkheit	mouldy	schimmlig; Schimmel. . .
alloy	(Metall)Legierung	pole vault	Stabhochsprung
ankle joint	Knöchelgelenk	racing track	Rennbahn
artificial	künstlich	resin	Harz
core	Kern	slope	Hang
efficiency	Tüchtigkeit, Leistungsfähigkeit, Wirkungsgrad	stamina	Ausdauer, Durchhaltevermögen
		supplement	ergänzen
hull	Rumpf	tendon	Sehne
laminate	(Plastik-, Verband) Folie	tread	Lauffläche
		varnish	Lack
leisure sports	Freizeitsport		

Questions

1) Why are sports events a competition of materials as much as one of athletes? Give reasons and examples.
2) How important are carbon fibres? What are their advantages and disadvantages?
3) What are the positive and negative effects of using chemicals in the athlete's diet?

Explain the following words:

a) competition
b) substitute
c) synthetic
d) sophisticated
e) resistance
f) hormone
g) artificial
h) vulnerable

Grammar

The Infinitive
without "to"

a) She can speak English.
   He shall analyse this substance.
   You may open the drying oven.
   – after the defective auxiliaries:
   can, shall, may, ...

b) I would rather leave now.
   The soldiers said they would sooner die than give in.
   Such a speech cannot but have had a great influence on the audience.
   They had better show their fares.
   – after the expressions:
   ....

c) She makes me cry.
   It would be a mistake to let slip such an opportunity.
   She would have them know that she cannot come.
   – after the verbs: to make,
   ....

d) She watched him wash the car.
   He felt his heart beat violently.
   We heard the bird sing.
   – after verbs of perception: to see,
   ....

Exercise

Translate into English:

1) Du tätest besser daran den Mund zu halten.
2) Der Polizist sah, wie der Gangster davonrannte.
3) So ein Mann muß einfach Erfolg haben.
4) Jeder muß im Labor eine Schutzbrille tragen.
5) Ich fühlte, wie meine Knie zitterten.

Beauty is Only Skin-Deep

As long ago as 3000 B. C. Egyptian women tried to outwit nature or, at least, tried to improve it. And for a long time, it was nature alone which provided man with cosmetics.

Artistic hair-styles (as well as wigs) were fixed with beeswax. A mixture of resins and beeswax was used to remove undesirable body-hair. Cheeks were tinged with ochre. Hands and fingernails were coloured with henna. Excavation finds brought eye make-up and fragrant oils to light.

Coins tell us of elaborate hair-styles of Roman women. They plaited blonde tresses from Germanic slaves and bleached their own hair with lime.

Generally, care of the body was at a premium. The preparation of a 'Cold Cream' was described by the physician Claudius Galenus, 200 A. D.:

Olive oil and beeswax are melted together. Then water is added as it cools. Some drops of rose-water complete the lotion.

The principle hasn't changed much since then: today's products, marketed under the label 'Cold Cream', have a cooling effect on the skin as water slowly evaporates from it.

Empress Poppaea never, not even while away, went without her usual bath in asses' milk. Today it is much easier for women (and men) to take a refreshing 'bubble bath', although highly complicated chemistry is involved (surface active substances reduce the surface tension thus making the bubbles last longer).

Solid soap, as we know it today, was unknown during the time of the ancients. To care for their bodies they used bran, pumice stone, oil, alumina, sodium, potash, vegetable alkali, soap wort, etc., which they often mixed together.

Sumerians and Egyptians boiled oils and fats with alkaline substances.

The Hittites dissolved vegetable ash in water to wash their hands.

Theokrit mentions a soap made from (animal) fat and ash.

Galen, the first to give an account of the usage of soap for cleansing the body and clothes, preferred the softer 'Germanic Soap'.

Later, better soaps were made by using quicklime, a by-product of the manufacture of ash-lye.

Today, soaps, manufactured as bars, granules, flakes, or in liquid form, are made from a mixture of the sodium salts of various fatty acids of natural oils and fats.

Shampoos:
their richness is rendered by the addition of sodium chloride, and stearic acid provides their nacreous lustre.

Sprays:
on abruptly opening the valve, a liquid propellant evaporates-atomizing the odorous substances present.

Hair:
smoothing out and curling hair can easily be done; a perm is a lasting alteration of a hair-style and a direct interference with the molecular structure of the hair substance (keratin fibres are held together by sulphuric atoms). After applying a solution of thioglycolate the hair loses its inner structure, it loosens and can be set in any style desired by applying an oxidizing agent.

Nail varnish:
colour and lustre are first assessed; then how well it can be applied and the drying time; moreover the varnish is supposed to adhere tightly and yet is expected to be hard and elastic at the same time. To meet all these needs, a balanced combination of solutions – softeners – is necessary: a resin providing adhesion and lustre, several kinds of nitrocellulose, and finally pigments to give the desired colour.

Lipsticks, tooth pastes, creams and perfumes are all products which also owe a great deal to chemistry. It has improved them and has made them versatile; in doing so it has provided them with the qualities which consumers demand today.

Lately, however, there has been a growing concern as to the harmful side effects of some substances:

propellants are said to destroy the vital ozone layer of the earth and hair dyes are suspected to be carcinogenic.

Chemistry and beauty:

Here science meets the demands of luxury and necessity. It is up to a society with a high sense of responsibility to apply stringent standards when it comes to deciding whether a substance may be used or not.

Now let's produce our own face lotion
A very simple face lotion can be made, when ⅔ water is added to ⅓ ethyl alcohol with a dash of glycerine (e. g. 65 ml of distilled water, 33 ml of ethyl alcohol, 2 ml of glycerine). To give this mixture a nice smell we can now add, for example, rose water or orange-blossom water.

Or how about making a moisturizing cream?
To do this we melt together
– liquid paraffin: 55% by volume,
– vaseline: 30% by volume, and
– beeswax: 15% by volume at about 85–90°C.

After cooling down to about 45°C, 5 drops of a sweet smelling substance are stirred in. Afterwards this mixture is poured into small Petri dishes for congelation.

Vocabulary

adhere	kleben, haften	potash	Pottasche
adhesion	Adhäsion	at a premium	hoch im Kurs
alumina	Tonerde		(stehend)
amalgamate	(sich) verschmelzen	propellant	Treibmittel
asses' milk	Eselsmilch	pumice stone	Bimsstein
bleach	bleichen	quicklime	gebrannter,
bran	Kleie		ungelöschter Kalk
congelation	Erstarren, Festwer-	render	verleihen; leisten;
	den, Gerinnen		ermöglichen
dye	Farbe	smooth	glätten
excavation	Ausgrabung	(out)	
flake	Flocke	soap wort	Seifenkraut
fragrant	wohlriechend	stearic acid	Stearinsäure
lye	Lauge	stringent	streng
nacreous	Perlmutt. . .; perl-	tension	Spannung
	muttartig	tinge	tönen, färben
odour	Geruch; Duft	tress	Locke; (Haar)
outwit	überlisten		Flechte, Zopf
perm	Dauerwelle	undesirable	unerwünscht
plait	(ver)flechten	wig	Perücke

Questions

1) What sorts of cosmetics did women use in the past?
2) How does present day soap differ from the soap used in the past?
3) How has chemistry helped to improve beauty products today?

Paraphrase or give synonyms for the following words:

a) cosmetics
b) lotion
c) refreshing
d) alkali
e) evaporate
f) render
g) lustre
h) adhesion
i) congelation

Grammar

Plural
Methods of forming the plural

1) dog, dogs; beaker, beakers, etc;   – add "-s"

2a) bench, benches; box, boxes, etc;   – add "-es"

....   after sibilant (hissing sound)

b) tomato, tomatoes; ....
                          after "o"
    but: piano, pianos, etc.

c) sky, skies; laboratory, laboratories, etc;

    ....   after changing "y" to "i"

    but: boy, boys, etc;   because ....

d) wolf, wolves, etc. ....   after changing "f" to "v"

    but: safes, reefs, etc. ....

3) child, children; ox, oxen; etc.   – add "en" or "ren"

4) man, men; tooth, teeth; etc.   – change a vowel

5) bacillus, bacilli; etc.   – a number of loan words have retained
    stratum, strata; etc.   their original plural forms
    nebula, nebulae, etc.
    phenomenon, phenomena; etc.
    analysis, analyses; etc.
    matrix, matrices; etc.
    bureau, bureaux; etc.

6) sheep, species, knowledge, infor-   – some nouns make no change for the
    mation, progress, etc.   plural

    ....

7) index   indices ....   – the two plurals have two quite distinct
                                                    meanings
               indexes ....

    genius   genii ....

               geniuses .... etc.

8) schoolmasters, spoilsports,   –
    passers-by, fathers-in-law,
    stowaways, teach-ins, etc.

    ....

9) Mathematics is my favourite   – sciences ending in -ics have the verb
    subject!   in the singular
    politics, mechanics, physics, etc.
    Bad news travels fast!

10a)   These scissors are no good.   – things consisting of two parts
        I have three pairs of scissors.
        glasses, trousers, etc.

  b)   The stairs were very steep.   – things consisting of several parts
        Finally we reached the outskirts
        of the city.
        vegetables, contents, etc.

        ....

                                                 Translate and explain

Note: Three pounds is enough.   – ....
           Fifty yards is a good distance.

           _____

           Every means has been tried.   – ....
           Measles is an infectious disease.

           _____

           The police are not interested in   – ....
           him. The police is . .
           The committee was/were
           appointed some time ago.

           _____

           A ten-penny piece; a six-cylin-   – ....
           der car; a two-room flat, etc.

           _____

           Many people lost their lives.   – ....
           They shook their heads.

           _____

Exercises

Translate the following sentences:

1) Hast du meine Brille gesehen? Du mußt sie im Büro gelassen haben.
2) Die Umgebung ist sehr schön.
3) Die Polizei hat die Diebe gefangen.
4) Mir taten die Füße weh.
5) Sie hat einige Fortschritte gemacht.
6) Statistik ist eines der wichtigsten Gebiete in der Mathematik.
7) Die Vereinigten Staaten sind ein großes Land.
8) Seine Hypothesen waren falsch.
9) Die Zuschauer waren begeistert.
10) Meine Schwester ließ sich die Haare schneiden.

Chemistry + Art = 2 Kinds of Art

General conceptions of what can be considered to be art have changed radically in this century. 100–200 years ago, the Bulgarian artist Christo, for instance, who drew a curtain across the Colorado Valley and declared this "art", would probably have been regarded as mad. But he would not have been able to exhibit his art show in the last century without the help of chemistry.

Chemistry has developed modern materials which many of today's artists experiment with.

The link between chemistry and art, however, has existed for a very long time: for many years clay, metal and colours have been basic materials of art, and chemistry is by definition the science dealing with matter, its characteristics and its transformation.

Painter's colours, old or new, are results of relatively complicated chemical combinations.
Red was made from cinnabar (mercury and sulphur-) – today, a less poisonous material, namely cadmium red, is used.

Blue was made from ultramarine (obtained from lapis lazuli), a substance whose combination was not discovered before this century.

Another old blue colour is indigo, which was extracted from plants. By the way, indigo is the standard colour for blue jeans.

When restoring old paintings art historians and restorers come up against problems which cannot be solved by a simple stylistic examination. Here, a chemical colour analysis is needed. In order to analyse colour combinations the chemist needs the same materials which the artist used to paint with. Therefore he takes small particles from the edge of a precious painting and puts them into an artificial resin. The resin block is cut and polished and the cross-section is then photographed under a microscope and analysed. The analysis informs him about the painting technique. Rubens, for example, was a "quick" artist, whereas other painters applied several layers of colour. However, not only the painting technique but also the chemical origins of the colours can be examined.

The apparatus, which separates light into different colours, is called a spectrograph. The chemical compounds of all metals – and most old paints are metal compounds – emit light if they are heated strongly enough (in an electric arc). Some of the light is not visible to the human eye, but nevertheless it can blacken a photographic plate.

On this plate lines are projected which correspond to a specific metal. The position of the lines indicates the type of metal, and the intensity of the lines shows the quantity of the metal present. This method of analysis is characterized by a tremendous sensitiveness.

Even more sensitive is the neutron-activating analysis, which is most appropriate for the analysis of ceruse. This colour was the only significant colour during the Middle Ages. Not until the middle of the 19th century did zinc white appear and at the beginning of the 20th century titanium white came into use. By this accurate method it is possible to say whether a given painting is an original or not.

Matter – subject of chemistry and the Fine Arts as well – disappears. It is reduced by micro-organisms and the oxidation of metallic particles. Here, chemistry can help to renovate faded paintings: the oxidation is reverted by hydrochloric acid and zinc dust.

The damage done by oxidation to old works of art in the course of centuries is done today by pollution in just 10–20 years. A similar process to the one mentioned above is applied for the restoration and preservation of antique statues made of bronze: the protective effects of the patina – a corrosional product of copper – are destroyed by sulphur combinations which are found in our polluted air. This sulphuric compound changes copper into copper sulphate which is washed off by the rain. Thus more and more precious metal is exposed to damage. Restorers and chemists clean the corroded surface without harming the metal and cover it with the synthetic material silicone. They hope that statues treated in this way can resist the city atmosphere forever.

Chemistry has been very successful in the conservation of historical buildings which are many centuries old but which have been exposed to rapid damage during the last decades. Most buildings and monuments are made of marble (limestone and sandstone). Although the sandstone grains contain quartz, which is indestructible, the binding agent – limestone – crumbles away. Air pollution reduces the limestone to plaster which ist finally washed off by rain. Thus the stone is exposed to gradual damage. Here again, the stone is cleaned. Then silicic acid is poured onto the surface. A chemical reaction produces siliconedioxide, which is a new, highly resistant and binding agent for the sandstone. Finally, silicone is applied to close the pores-thus preventing polluted water from soaking in and causing further damage.

Large buildings, like "Die alte Pinakothek" in Munich or the Cathedral in Cologne, cause temporal problems for the restorers: their time to restore and protect large objects is running short. Therefore smaller originals are often exhibited in museums and reproductions are placed in the old spot. Synthetic reproductions are hard to differentiate from their originals.

The fact that it is often difficult to distinguish the original from the reproduction arouses the artist's suspicion. To his mind the original material is an essential part of what the artist wanted to express. Are not expression, form and material one entity that should not be separated?

Vocabulary

arc	Bogen	entity	Wesen; Einheit
cast	gießen, formen; gestalten	exhibit	ausstellen
		extract	(her)ausziehen, herauslösen
ceruse	Bleiweiß		
cinnabar	Zinnober	fade	verblassen; verbleichen
clay	Lehm, Ton		
conception	Vorstellungskraft; Auffassung, Idee, Gedanke; Plan	indestructible	unzerstörbar
		lapis lazuli	Lapislazuli
		link	(Ketten) Glied; (Binde) Glied, Verbindung
correspond	entsprechen		
cross-section	Querschnitt		
crumble	zerkrümeln; zerbröckeln, zerfallen	plaster	(Ver) Putz; Gips
		polish	polieren, blank reiben; glätten
decade	Jahrzehnt		
differentiate	(sich) unterscheiden, trennen	suspicion	Verdacht, Argwohn
		tremendous	gewaltig, riesig; toll hervorragend
distinguish	unterscheiden; auseinanderhalten		

Questions

1) How has chemistry helped to develop art materials during the centuries?
2) How can chemistry aid art historians in their work? (With special reference to the spectrograph and the neutron-activating analysis.)
3) Name and describe some of the ways in which chemistry has been used to conserve historical buildings?

Paraphrase or give synonyms for the following words:

a) conception
b) transformation
c) cast
d) chemical compound
e) characterize
f) appropriate
g) oxidation
h) preservation
i) reduce
j) exhibit

Grammar

The Use of Capital Letters

Capitalize   **proper nouns,**
               proper adjectives, and their
               abbreviations.

A proper noun is the name of a particular person or thing. A common noun refers to any one of a class of persons or things and is not capitalized.

Proper Noun	Common Noun
Claudius Galenus	a physician
Empress Poppaea	an empress
Ohio	a state
Queen Elizabeth	a queen

The names of school subjects except languages are common nouns:
          algebra, chemistry, history etc.
But: Latin, German, English, etc.

Proper adjectives are derived from proper nouns:
          England – English history
          Christ – Christian principles
          Egypt – Egyptian women

Some adjectives derived from proper names are not capitalized:
          china vase, pasteurized milk, puritanical character, roman type, etc.

Proper nouns include:

1) Titles of organizations and institutions:

Proper	Common
Harvard University	a university
Chemical Institute of Dr. Fleet	an institute
Grace Episcopal Church	a church

2) Days of the week, months of the year, and holidays (but not seasons):
        Monday, April, Easter, etc.
   But: summer, autumn, spring, etc.

3) Geographical names:
        River Thames, The Rocky Mountains, Yellow Stone Park, etc.

4) Names of buildings:
        Big Ben, Empire State Building, Eiffel Tower, etc.

5) Words like North, East and Northwest when they name particular parts of the country.
When these words refer to directions, they are not capitalized:
        The Midwest of the USA is one of the greatest food-producing areas in the world. (part of country)
        Princeton is eleven miles north of Trenton and thirteen miles south of Somerville. (direction)

6) Names of political parties, religious sects, nations, and races:
        Republican, Catholic, Negro, Japanese, etc.

7) Historical events, periods, and documents:
        Declaration of Independence, Stone Age, Middle Ages, Treaty of Versailles, etc.

8) Names of governmental bodies and departments:
        Congress, Senate, The Board of Education, etc.

9) Titles before proper names and titles of high government officials used without proper names.
        President Reagan, Governor Wallace, etc.
   But: A senator and a governor were speakers at the meeting.
        Capitalize **Aunt, Cousin,** and **Uncle** before proper names: Aunt Mary and Uncle Bill will visit us in December.

10) The names of the planets:
        Jupiter, Venus, Mercury, Earth, etc.

11) The specific part of the trade name of a product:
        Ivory soap, Pioneer radio, etc.

12) Titles of books, articles, poems, and newspapers.
        In titles the first and the last word and other words except articles, prepositions, and conjunctions are capitalized.
        Three Men in a Boat, Beauty is only Skin-deep, The Wreck of the Hesperus, the New York Times, etc. (Do not capitalize **the** as the first word of the name of a newspaper or magazine)!
        the Reader's Digest, etc.

13) Names referring to the Deity, the Bible, and divisions of the Bible:
        the Almighty, the Lord, the Scriptures, Revelations, etc.

Exercise:

In each sentence and expression capitalize the words that need capitalization. Give the reason for every capital letter.

1) Cousin david, who is a student at the university of heidelberg, studies english, chemistry, mathematics, history, and science.
2) Charles dickens, one of england's great writers, is the author of **david copperfield** and **a tale of the two cities.**
3) The planet mars has been the subject of much speculation by scientists and by such writers as h. g. wells.
4) John eliot, who was educated at cambridge university and came to america in 1631, translated the bible into the language of the indians.
5) On christmas night washington crossed the delaware river and at daybreak surprised the force of hessian soldiers which general Howe had stationed at trenton.
6) wednesday evening / french food / the society of plastics industry.

Bacchus in the Laboratory

The science examining the chemical principles of vital functions is called BIOCHEMISTRY. In 1897, the history of this science, which was still in its infancy, encountered a striking incident when Eduard Buchner, a chemist, discovered that fruit juice could ferment into alcohol without the addition of living yeast. This process also works if yeast is mixed with quartzsand and diatomaceous earth. The resulting juice is pressed and filtered into an extract containing no living matter. Buchner filled one vat containing grape juice with this extract, another one with yeast of wine. The fermentation took place in both vats just the same. Thus the experiment showed that vital processes are complicated but normal **chemical** processes without any magical power.

Yet, even today we cannot neglect active yeast in order to produce a special group of chemical substances: FERMENTS in former days, ENZYMES as they are called today, accelerate and regulate chemical processes.

Especially with alcoholic fermentation they are of major importance for a series of reactions. Connoisseurs of wine, however, are satisfied when told of the original material and the final product: carbohydrates from grapes are changed into alcohol and carbon-dioxide (CO_2). Yet, even the most serious eco-supporter has to know that although no chemical substances is added in the development of wine, it is a multitude of chemical substances and reactions that gives the final product its flavour and bouquet.

From the beginning and throughout its growth, the grape is in constant contact with chemistry. Pesticides are applied to prevent it from damage, such as fungus. In the following process chemistry is essential, too, although chemical interference is strictly limited within the European Common Market. Wine is by far the most strictly controlled food product. The grapes are trodden and without any other processes we would get just white wine. Red wine is a must without stalks, parts of which could have negative effects on the flavour of the wine. In former times the must of red grapes was fermented directly to obtain red wine. Today the must is heated briefly thus allowing the skin's red colour to pass into the juice. Then the vats to contain the wine are prepared with sulphur dioxide to kill harmful micro-organisms which could affect the wine's bouquet and colour.

Large vine areas in Germany do not get sufficient sun for the growth of quality wines; their grapes have too much acid and too little sugar. Here we find a limited addition of sugar and of lime to reduce the concentration of wine-acid. Cabinet wines, however, are banned from any addition of sugar or lime.

Another legalized process used with quality wines is the sterilisation at about 85°C which keeps the grape juice from moulding. Later this specially treated juice is added to the rest and produces a milder bouquet.

Throughout all the stages mentioned above the must has never been a clear juice. All the sediment is now removed by filtration and centrifugal force, then yeast is added to the purified juice and the fermentation begins. The escaping gas from carbon dioxide indicates the process, whose first result is a murky juice called "Federweisse" (cider).

After several weeks the young wine is separated from the sediment. Again sulphur dioxide has to be applied to sterilize the wine and remove acetaldehyde – an unhealthy chemical compound. Whereas iron and cooper particles had to be removed in earlier times, these problems are not so great now-thanks to vats made of stainless steel or plastic. Only a few fermentation substances have to be eliminated. The final product is kept in vats for several months – sometimes years – and then it is poured into bottles and delivered to the consumer.

Chemists are sometimes confronted with tampered wine when the results in the vineyards are below average because of bad weather in spring or heavy damage by thunderstorms in summer or autumn. Water and sugar often help to improve the quality of the wine. If used during the correct stage of the wine's development and in moderate quantities the addition of sugar is allowed, as we have seen. Yet, some tamperers try to sell poor quality wines as high quality wines so as to get a good profit. To improve the flavour they use glycerine – a natural by-product of wine fermentation. It is this fact that makes it rather difficult for the chemist to analyse whether the glycerine proportion is in an adequate quantity. But he can prove tampering by comparing his analysis with the average results.

Thus chemistry is essential for the improvement of aromatic elements and the cultivation of new wines. Hence the vinedresser finds out the best date to start the annual wine harvest.

Here's to you!

Determination of Alcoholic Strength

There are several ways to determine the concentration of alcohol in wine, for example, by means of

– distillation and determination of the density in the distillate,
– the density,

– the determination of the refractive index (nomograms), and
– titrations, for example, with potassium bichromate.

In the following experiment we want to determine the concentration of the alcohol present.

Establish the density of the given solution consisting of water and alcohol and then determine the concentration of the latter by referring to the chart below.

Density (at 20°C)	Gr. of Alcohol/Liter									
0.999	0.5	1.1	1.6	2.1	2.7	3.2	3.7	4.3	4.8	5.3
0.998	5.8	6.4	6.9	7.4	8.0	8.5	9.0	9.6	10.1	10.6
0.997	11.2	11.7	12.3	12.8	13.4	13.9	14.5	15.0	15.5	16.1
0.996	16.6	17.2	17.7	18.3	18.8	19.4	19.9	20.5	21.0	21.6
0.995	22.1	22.7	23.3	23.8	24.4	24.9	25.5	26.1	26.6	27.2
0.994	27.8	28.3	28.9	29.4	30.0	30.6	31.2	31.8	32.4	32.9
0.993	33.5	34.1	34.7	35.3	35.9	36.5	37.1	37.6	38.2	38.8
0.992	39.4	39.9	40.5	41.1	41.7	42.3	42.9	43.5	44.1	44.7
0.991	45.3	45.9	46.5	47.1	47.7	48.3	48.9	49.5	50.1	50.7
0.990	51.3	52.0	52.6	53.2	53.8	54.4	55.0	55.6	56.2	56.9
0.989	57.5	58.1	58.7	59.3	59.9	60.6	61.2	61.8	62.5	63.1
0.988	63.7	64.4	65.0	65.6	66.3	66.9	67.5	68.2	68.8	69.4
0.987	70.1	70.7	71.4	72.0	72.7	73.3	74.0	74.6	75.3	75.9
0.986	76.6	77.2	77.9	78.5	79.1	79.8	80.4	81.1	81.8	82.5
0.985	83.1	83.8	84.5	85.1	85.8	86.5	87.2	87.8	88.5	89.2
0.984	89.9	90.6	91.2	91.9	92.6	93.3	94.0	94.7	95.4	96.0
0.983	96.7	97.4	98.1	98.8	99.5	100.2	100.9	101.6	102.3	103.0
0.982	103.6	104.3	105.0	105.7	106.4	107.1	107.8	108.5	109.2	109.9
0.981	110.7	111.4	112.1	112.8	113.5	114.2	114.9	115.7	116.4	117.1
0.980	117.8	118.5	119.3	120.0	120.7	121.5	122.2	122.9	123.6	124.4
0.979	125.1	125.8	126.6	127.3	128.0	128.8	129.5	130.2	130.9	131.6
0.978	132.4	133.1	133.8	134.5	135.3	136.0	136.7	137.4	138.2	138.9
0.977	139.7	140.4	141.2	141.9	142.7	143.4	144.2	144.9	145.7	146.4
0.976	147.1	147.9	148.6	149.3	150.1	150.8	151.5	152.2	153.0	153.7
0.975	154.4	155.2	155.9	156.6	157.4	158.1	158.8	159.6	160.3	161.0
0.974	161.7	162.5	163.2	163.9	164,7	165.4	166.1	166.9	167.6	168.3
0.973	169.1	169.8	170.5	171.2	172.0	172.7	173.4	174.2	174.9	175.6
0.972	176.3	177.0	177.8	178.5	179.2	179.9	180.6	181.3	182.1	182.8
0.971	183.5	184.2	184.9	185.6	186.3	187.0	187.7	188.4	189.1	189.8
0.970	190.5	191.2	191.9	192.6	193.3	194.0	194.7	195.4	196.1	196.8
0.969	197.5	198.2	198.8	199.5	200.2	200.9	201.6	202.3	203.0	203.7
0.968	204.0	205.0	205.7	206.4	207.0	207.7	208.4	209.0	209.7	210.4
0.967	211.1	211.7	212.4	213.1	213.7	214.4	215.1	215.7	216.4	217.1
0.966	217.7	218.4	219.0	219.7	220.4	221.0	221.7	222.3	223.0	223.6
0.965	224.3	224.9	225.6	226.2	226.9	227.5	228.2	228.8	229.5	230.1

Vocabulary

adequate	angemessen, ausreichend	the latter	(der, die das) letztere
bouquet	Bukett, Blume (des Weines)	moderate	gemäßigt; mäßig; angemessen
chart	Schaubild, graph. Darstellung; Tabelle	multitude	Menge; Masse
diatomaceous	Diatomeen...	murky	trübe; dunkel
diatomaceous earth	Kieselgur	must	Most
		neglect	vernachlässigen; versäumen, unterlassen
encounter	treffen; stoßen auf	nomogram	Nomogramm
ferment	gären lassen; (ver)gären	refractive	Brechungs..., Refraktions...
flavour	Geschmack, Aroma	stainless	flecken-, makellos; rostfrei
hence	also, folglich, deshalb	stalk	Stengel, Halm
incident	einfallend; vorkommend; zusammenhängend	tamper	panschen, verfälschen
infancy	Anfangsstadium	vat	Faß, Bottich
interference	Einmischung; Störung	vinedresser	Winzer

Questions

1) What was Buchner's contribution to the study of biochemistry?
2) What role does 'living yeast' play in chemical processes?
3) Describe briefly the ways in which chemistry is used in the making of wine.
4) How and why is chemistry used to tamper with wine?

Find synonyms or paraphrases for the following words:

a) encounter   b) filter   c) fermentation   d) accelerate
e) micro-organism   f) confront   g) tamper

How to read: Chemical Formulae

NaCl /ˌen eɪ siː ˈel/
$2H_2 + O_2 = 2H_2O$ /ˌtuː eɪtʃ ˌtuː‖ plʌs ˌəʊ ˈtuː‖ ˌiːkwəlz ˌtuː eɪtʃ tuː ˈəʊ/

NaOH   ˌen eɪ ˈəʊ eɪtʃ
$Na_2CO_3 + H_2CO_3 \longrightarrow 2NaHCO_3$ ˌen eɪ tuː siː ˈəʊ θriː ‖ plʌs eɪtʃ ˌtuː siː ˌəʊ ˈθriː ˌiːkwəlz ˌtuː ˌen eɪ eɪtʃ siː ˌəʊ ˈθriː

> **Note:** In chemical formulae capital and lower case letters, and full-size and reduced-size numbers **are not distinguished orally.**

How to read: Mathematical and Physical Expressions/Symbols

$a = b$	a equals b; a is equal to b
$a \neq b$	a is not equal to b
$a > b$	a is greater than b
$a < b$	a is less than b
$a \parallel b$	a is parallel to b
$a + b$	a plus b
$a - b$	a minus b
$a \pm b$	a plus or minus b
$a \times b$	a times b; a multiplied by b
$a \div b$	a divided by b
$\dfrac{a}{b}$	a over b
$a'b''$	a dash times b double dash
$(a); [a]; \{a\}$	a in parentheses; a in brackets; a in braces
$a(b + c)$	a parenthesis open b plus c parenthesis close
$\sin x$	sine x
$\cos x$	cosine x
$(\sin x)' = \cos x$	the (first) derivative of the sine of x equals the cosine of x
$\dfrac{dy}{dx}$	dy by dx
$f(x)$	function of x; f of x
$\int$	integral
10^1	ten to the power of one; ten to the power one; ten to the first power; ten to the first
10^2	ten to the second; ten squared
10^∞	ten to the power of infinity
10^{-1}	ten to the minus first power
a^{-3n}	a to the power of minus 3 n
$\sqrt{a}$	square root of a
$\sqrt[3]{a}$	third (cube) root of a
$\ln x$	natural logarithm of x
$\log x$	common logarithm of x
$\log_{10} 2 = 0.30103$	logarithm of two to the base ten is nought point three, nought, one, nought, three
$\frac{1}{2}$	a (one) half
$\frac{1}{3}$	a (one) third
$\frac{1}{4}$	a (one) quarter, a (one) fourth
$\frac{1}{5}$	a (one) fifth
$7\frac{4}{5}$	seven and four fifths
0.37	nought point (decimal) thirty-seven; (or: three seven)
∞	infinity

Now practise

11^{-7}; $90 - 30 = 3$; 9%; $3.5 \times 4.73 = 16.555$; $(7 - 4)\, 3 =$ ....

$Zn + H_2SO_4 \rightarrow ZnSO_4 + H_2$; ....

$Ca(OH)_2 + CO_2 \rightarrow CaCO_3 + H_2O$; ....

Complete the following table:

sign	noun	verb
+	addition	add
−		
×		
÷		

Tricks and Magic

"Bend!! Break!! Bend!! Break!!"

Uri Geller is at work. Will he succeed?

For many onlookers the question arises whether this man is a sham or not. At least he seems to belong to the unpleasant category of conjurers that appeal to the superstitious instincts of his audience. But, it has to be said, his almost perfect sleight-of-hand is amazing.

Comparatively honest tricks are shown by a second type of conjurers: the professional magicians and illusionists. And thirdly there are scientists performing magic. One feels that everything must have a plausible explanation – but how can it be explained? Sometimes they are the ones who really seem to be able to work miracles.

Now let's turn our attention to some of the 'tricks' they can perform!

– Two halves of a ball are tightly screwed together, heated and then an invisible gas is ignited. The resulting jet of flame can be as much as 12 metres high (Imhoff-Sphere).

Illusionists have always made use of coloured, spectacular flames. Today it is possible to vary the colour and brightness of a rocket by selecting proper fuels and metal additives: white rays contain metal powder such as magnesium or aluminium, intensive colours contain plastics as fuel and a good deal of metallic salt.

– A rubber hose is put into a tank filled with liquid. Then it is taken out and smashed to pieces.
A rose is dipped into the same liquid, taken out again and crumbled between the fingers.

What magic liquid was used?

– One liquid is carefully poured onto another, without them mixing. Then with a pair of tweezers, a thread is pulled from the dividing line of the two liquids. In this way a thread of more than 10 metres can be pulled out. In chemists' circles this experiment is known as the 'Nylon-Rope-Trick'.

Guess what liquids must be used!

Even more impressive and scientifically baffling is the following performance where a change of colour takes place after a given time.

Now let's get down to making an

'Oscillating Iodine Clock'

To 1 l of water add these substances in the following order:

– 14.3 g of potassium iodate,
– 122.4 ml of hydrogen peroxide (30%),
– 5.9 ml of perchloric acid (60%),
– 5.2 g of malonic acid,
– 10 ml of starch (1%),
– 1.1 g of manganese(II) sulfate.

The change of colour begins when manganese is added.

Vocabulary

appeal to	ansprechen, (sich) wenden an; wirken auf, zusagen	rubber hose	Gummischlauch
		sham	Heuchler, Scharlatan
baffling	verwirrend	sleight-of-hand	Kunststück, (Ta-schenspieler-)Trick
conjurer	Zauberkünstler, Zauberer	superstitious	abergläubisch
ignite	(sich) entzünden, anzünden	tweezers	Pinzette

Questions

1) What are the three types of magicians?
2) How do they differ?
3) Name some of the ways chemistry produces tricks.

Explain the following words:

a) superstitious   b) plausible   c) instinct
d) spectacular   e) crumble   f) baffling
g) performance

Grammar

Defining and non-defining relative clauses

1. People
who have plenty to do
are seldom bored.

The pencil
that is lying on the table
is a very good one.

2. His eldest brother,
who lives in Hamburg,
is seriously ill.

The amplifiers,
**which form part of the link between
the source and the listener,**
can have a marked effect on the quality
of the reproduction.

Compare the defining relative clauses of 1 with the non-defining relative clauses of 2 and point out their differences.
What do we understanding by 'defining' and 'non-defining'?

Exercise

Insert the correct relative pronouns and decide whether to put commas or not:

1) It is the electrical engineer ....designs and constructs electrical devices.

2) My English in .... I failed most exams was always poor.

3) It is the output stage .... should provide sufficient power to guarantee realistic reproduction of orchestral music.

4) The teacher .... taught us English was called Mr Bailiff.

5) It is the amplifier .... is the only means of control or adjustment in any Hi-Fi equipment.

6) Latin .... was my second foreign language was not my favourite subject.

7) This is a reproduction .... a discerning listener can be satisfied with.

8) She is a women in .... we can always put our trust.

9) Her methods include everything .... makes learning easy.

5 Appendix

Individual Sounds
The Greek Alphabet
Relative Strengths of Acids
Periodic Table
Relative Atomic Masses (Atomic Weights)
Numerical Expressions
Weights and Measures
Irregular Verbs

Individual Sounds[4]

Phonetic Symbol	Example	Phonetic Symbol	Example
iː	green	p	pen
ɪ	sit	b	ball
e	yes	m	man
æ	cat	w	we
ɑː	glass	f	full
ɔ	hot	v	very
ɔː	ball	θ	thin
ʊ	book	ð	the
uː	soon	t	ten
ʌ	cup	d	do
ɜː	bird	l	live
ə	again	n	no
eɪ	day	r	red
əʊ	boat	s	sit
aɪ	buy	z	zoo
aʊ	cow	ʃ	shut
ɔɪ	boy	ʒ	measure
ɪə	fear	tʃ	chop
eə	chair	dʒ	jump
ʊə	fewer	j	yes
		k	car
		g	go
		ŋ	sing
		h	hop

The Greek Alphabet

Capitals	Small letters	Name	Usually designating
A	α	Alpha	Angles; coefficients . . .
B	β	Beta	Angles; flux density . . .
Γ	γ	Gamma	Angles; conductivity . . .
Δ	δ	Delta	Variation; density . . .
E	ε	Epsilon	Base of natural logarithms . . .
Z	ζ	Zeta	Impedance; . . .
H	η	Eta	Hysteresis coefficient; . . .
Θ	ϑ	Theta	Temperature; Time constant; . . .
I	ι	Iota	Current in amperes
K	ϰ	Kappa	Dielectric constant; Susceptibility; Visibility
Λ	λ	Lambda	Wave length; . . .
M	μ	Mu	Amplification factor; . . .
N	ν	Nu	Reluctivity
Ξ	ξ	Xi	Amplitude; function variable
O	o	Omicron	
Π	π	Pi	Ratio of circumference to diameter = 3.1416
P	ϱ	Rho	Resistivity
Σ	σ	Sigma	Sign of summation; . . .
T	τ	Tau	Time constant; . . .
Y	υ	Ypsilon	
Φ	φ	Phi	Angles; magnetic flux
X	χ	Chi	Reactance; . . .
Ψ	ψ	Psi	Phase difference; dielectric flux; . . .
Ω	ω	Omega	Angular velocity; Ohms; . . .

Relative Strengths of Acids[1]

In Aqueous solution at room temperature

All ions are aquated

$$HB \rightleftarrows H^+(aq) + B^-(aq) \qquad K_A = \frac{[H^+][B^-]}{[HB]}$$

Acid	Strength	Reaction	K_A
perchloric acid	very strong	$HClO_4 \rightarrow H^+ + ClO_4^-$	very large
hydriodic acid		$HI \rightarrow H^+ + I^-$	very large
hydrobromic acid		$Hbr \rightarrow H^+ + Br^-$	very large
hydrochloric acid		$HCl \rightarrow H^+ + Cl^-$	very large
nitric acid		$HNO_3 \rightarrow H^+ + NO_3^-$	very large
sulfuric acid	very strong	$H_2SO_4 \rightarrow H^+ + HSO_4^-$	large
oxalic acid		$HOOCCOOH \rightarrow H^+ + HOOCCOO^-$	5.4×10^{-2}
sulfurous acid ($SO_2 + H_2O$)		$H_2SO_3 \rightarrow H^+ + HSO_3^-$	1.7×10^{-2}
hydrogen sulfate ion	strong	$HSO_4^- \rightarrow H^+ + SO_4^{-2}$	1.3×10^{-2}
phosphoric acid		$H_3PO_4 \rightarrow H^+ + H_2PO_4^-$	7.1×10^{-3}
ferric ion		$Fe(H_2O)_6^{+3} \rightarrow H^+ + Fe(H_2O)_5(OH)^{+2}$	6.0×10^{-3}
hydrogen telluride		$H_2Te \rightarrow H^+ + HTe^-$	2.3×10^{-3}
hydrofluoric acid	weak	$HF \rightarrow H^+ + F^-$	6.7×10^{-4}
nitrous acid		$HNO_2 \rightarrow H^+ + NO_2^-$	5.1×10^{-4}
hydrogen selenide		$H_2Se \rightarrow H^+ + HSe^-$	1.7×10^{-4}
chromic ion		$Cr(H_2O)_6^{+3} \rightarrow H^+ + Cr(H_2O)_5(OH)^{+2}$	1.5×10^{-4}
benzoic acid		$C_6H_5COOH \rightarrow H^+ + C_6H_5COO^-$	6.6×10^{-5}
hydrogen oxalate ion		$HOOCCOO^- \rightarrow H^+ + OOCCOO^{-2}$	5.4×10^{-5}
acetic acid	weak	$CH_3COOH \rightarrow H^+ + CH_3COO^-$	1.8×10^{-5}
aluminum ion		$Al(H_2O)_6^{+3} \rightarrow H^+ + Al(H_2O)_5(OH)^{+2}$	1.4×10^{-5}
carbonic acid ($CO_2 + H_2O$)		$H_2CO_3 \rightarrow H^+ + HCO_3^-$	4.4×10^{-7}
hydrogen sulfide		$H_2S \rightarrow H^+ + HS^-$	1.0×10^{-7}
dihydrogen phosphate ion		$H_2PO_4^- \rightarrow H^+ + HPO_4^{-2}$	6.3×10^{-8}
hydrogen sulfite ion		$HSO_3^- \rightarrow H^+ + SO_3^{-2}$	6.2×10^{-8}
ammonium ion	weak	$NH_4^+ \rightarrow H^+ + NH_3$	5.7×10^{-10}
hydrogen carbonate ion		$HCO_3^- \rightarrow H^+ + CO_3^{-2}$	4.7×10^{-11}
hydrogen telluride ion		$HTe^- \rightarrow H^+ + Te^{-2}$	1.0×10^{-11}
hydrogen peroxide	very weak	$H_2O_2 \rightarrow H^+ + HO_2^-$	2.4×10^{-12}
monohydrogen phosphate ion		$HPO_4^{-2} \rightarrow H^+ + PO_4^{-3}$	4.4×10^{-13}
hydrogen sulfide ion		$HS^- \rightarrow H^+ + S^{-2}$	1.3×10^{-13}
water		$H_2O \rightarrow H^+ + OH^- \quad [H^+][OH^-] = 1.0 \times 10^{-14}$	
hydroxide ion		$OH^- \rightarrow H^+ + O^{-2}$	$< 10^{-36}$
ammonia	very weak	$NH_3 \rightarrow H^+ + NH_2^-$	very small

Periodic Table[2]

Group Period	I	II	Transition Metals										III	IV	V	VI	VII	O
1	1 H																	2 He
2	3 Li	4 Be											5 B	6 C	7 N	8 O	9 F	10 Ne
3	11 Na	12 Mg											13 Al	14 Si	15 P	16 S	17 Cl	18 Ar
4	19 K	20 Ca	21 Sc	22 Ti	23 V	24 Cr	25 Mn	26 Fe	27 Co	28 Ni	29 Cu	30 Zn	31 Ga	32 Ge	33 As	34 Se	35 Br	36 Kr
5	37 Rb	38 Sr	39 Y	40 Zr	41 Nb	42 Mo	43 Tc	44 Ru	45 Rh	46 Pd	47 Ag	48 Cd	49 In	50 Sn	51 Sb	52 Te	53 I	54 Xe
6	55 Cs	56 Ba	57 La	72 Hf	73 Ta	74 W	75 Re	76 Os	77 Ir	78 Pt	79 Au	80 Hg	81 Tl	82 Pb	83 Bi	84 Po	85 At	86 Rn
7	87 Fr	88 Ra	89 Ac															

Transition Metals

58 Ce	59 Pr	60 Nd	61 Pm	62 Sm	63 Eu	64 Gd	65 Tb	66 Dy	67 Ho	68 Er	69 Tm	70 Yb	71 Lu
90 Th	91 Pa	92 U	93 Np	94 Pu	95 Am	96 Cm	97 Bk	98 Cf	99 Es	100 Fm	101 Md	102 No	103 Lw

Relative Atomic Masses (Atomic Weights)[2]

Element	Symbol	Atomic number	Relative atomic mass	Element	Symbol	Atomic number	Relative atomic mass
Actinium	Ac	89	227.0	Mercury	Hg	80	200.6
Aluminium	Al	13	26.9	Molybdenum	Mo	42	95.9
Americium	Am	95	243.0	Neodymium	Nd	60	144.2
Antimony	Sb	51	121.8	Neon	Ne	10	20.2
Argon	Ar	18	39.9	Neptunium	Np	93	237.0
Astatine	At	85	210.0	Nickel	Ni	28	58.7
Arsenic	As	33	74.9	Niobium	Nb	41	92.9
Barium	Ba	56	137.3	Nitrogen	N	7	14.0
Berkelium	Bk	97	249.0	Osmium	Os	76	190.2
Beryllium	Be	4	9.0	Oxygen	O	8	16.0
Bismuth	Bi	83	209.0	Palladium	Pd	46	106.4
Boron	B	5	10.8	Phosphorus	P	15	31.0
Bromine	Br	35	79.9	Platinum	Pt	78	195.1
Cadmium	Cd	48	112.4	Plutonium	Pu	94	242.0
Caesium	Cs	55	132.9	Polonium	Po	84	210.0
Calcium	Ca	20	40.1	Potassium	K	19	39.1
Californium	Cf	98	251.0	Praseodymium	Pr	59	140.9
Carbon	C	6	12.0	Promethium	Pm	61	145.0
Cerium	Ce	58	140.1	Protactinium	Pa	91	231.0
Chlorine	Cl	17	35.5	Radium	Ra	88	226.1
Chromium	Cr	24	52.0	Radon	Rn	86	222.0
Cobalt	Co	27	58.9	Rhenium	Re	75	186.2
Copper	Cu	29	63.5	Rhodium	Rh	45	102.9
Curium	Cm	96	247.0	Rubidium	Rb	37	85.5
Dysprosium	Dy	66	162.5	Ruthenium	Ru	44	101.1
Einsteinium	Es	99	254.0	Samarium	Sm	62	150.4
Erbium	Er	68	167.3	Scandium	Sc	21	45.0
Europium	Eu	63	152.0	Selenium	Se	34	79.0
Fermium	Fm	100	253.0	Silicon	Si	14	28.1
Fluorine	F	9	19.0	Silver	Ag	47	107.9
Francium	Fr	87	223.0	Sodium	Na	11	23.0
Gadolinium	Gd	64	157.3	Strontium	Sr	38	87.6
Gallium	Ga	31	69.7	Sulphur	S	16	32.1
Germanium	Ge	32	72.6	Tantalum	Ta	73	180.9
Gold	Au	79	197.0	Technetium	Tc	43	99.0
Hafnium	Hf	72	178.5	Tellurium	Te	52	127.6
Helium	He	2	4.0	Terbium	Tb	65	158.9
Holmium	Ho	67	164.9	Thallium	Tl	81	204.4
Hydrogen	H	1	1.0	Thorium	Th	90	232.0
Indium	In	49	114.8	Thulium	Tm	69	168.9
Iodine	I	53	126.9	Tin	Sn	50	118.7
Iridium	Ir	77	192.2	Titanium	Ti	22	47.9
Iron	Fe	26	55.8	Tungsten	W	74	183.9
Krypton	Kr	36	83.8	Uranium	U	92	238.0
Lanthanum	La	57	138.9	Vanadium	V	23	50.9
Lead	Pb	82	207.2	Xenon	Xe	54	131.3
Lithium	Li	3	6.9	Ytterbium	Yb	70	173.0
Lutetium	Lu	71	175.0	Yttrium	Y	39	88.9
Magnesium	Mg	12	24.3	Zinc	Zn	30	65.4
Manganese	Mn	25	54.9	Zirconium	Zr	40	91.2

Numerical Expressions[3]

The following section will give you help in the reading, speaking and writing of numbers and expressions which commonly contain numbers.

Note. 'a /ə/ hundred' is a less formal usage than 'one /wʌn/ hundred'.

Cardinal	Ordinal
1 one /wʌn/	1st first /fɜːst/
2 two /tuː/	2nd second /'sekənd/
3 three /θriː/	3rd third /θɜːd/
4 four /fɔː(r)/	4th fourth /fɔːθ/
5 five /faɪv/	5th fifth /fɪfθ/
6 six /sɪks/	6th sixth /sɪksθ/
7 seven /'sevn/	7th seventh /'sevnθ/
8 eight /eɪt/	8th eighth /eɪtθ/
9 nine /naɪn/	9th ninth /naɪnθ/
10 ten /ten/	10th tenth /tenθ/
11 eleven /ɪ'levn/	11th eleventh /ɪ'levnθ/
12 twelve /twelv/	12th twelfth /twelfθ/
13 thirteen /ˌθɜː'tiːn/	13th thirteenth /ˌθɜː'tiːnθ/
14 fourteen /ˌfɔː'tiːn/	14th fourteenth /ˌfɔː'tiːnθ/
15 fifteen /ˌfɪf'tiːn/	15th fifteenth /ˌfɪf'tiːnθ/
16 sixteen /ˌsɪk'stiːn/	16th sixteenth /ˌsɪk'stiːnθ/
17 seventeen /ˌsevn'tiːn/	17th seventeenth /ˌsevn'tiːnθ/
18 eighteen /ˌeɪ'tiːn/	18th eighteenth /eɪ'tiːnθ/
19 nineteen /ˌnaɪn'tiːn/	19th nineteenth /ˌnaɪn'tiːnθ/
20 twenty /'twentɪ/	20th twentieth /'twentɪəθ/
21 twenty-one /ˌtwentɪ'wʌn/	21st twenty-first /ˌtwentɪ'fɜːst/
22 twenty-two /ˌtwentɪ'tuː/	22nd twenty-second /ˌtwentɪ'sekənd/
23 twenty-three /ˌtwentɪ'θriː/	23rd twenty-third /ˌtwentɪ'θɜːd/
30 thirty /'θɜːtɪ/	30th thirtieth /'θɜːtɪəθ/
38 thirty-eight /ˌθɜːtɪ'eɪt/	38th thirty-eighth /ˌθɜːtɪ'eɪtθ/
40 forty /'fɔːtɪ/	40th fortieth /'fɔːtɪəθ/
50 fifty /'fɪftɪ/	50th fiftieth /'fɪftɪəθ/
60 sixty /'sɪkstɪ/	60th sixtieth /'sɪkstɪəθ/
70 seventy /'sevntɪ/	70th seventieth /'sevntɪəθ/
80 eighty /'eɪtɪ/	80th eightieth /'eɪtɪəθ/
90 ninety /'naɪntɪ/	90th ninetieth /'naɪntɪəθ/
100 a/one hundred /ə, wʌn 'hʌndrəd/	100th a/one hundredth /ə, wʌn 'hʌndrədθ/
1000 a/one thousand /ə, wʌn 'θaʊznd/	1000th a/one thousandth /ə, wʌn 'θaʊznθ/
10 000 ten thousand /ˌten 'θaʊznd/	10 000th ten thousandth /ˌten 'θaʊznθ/
100 000 a/one hundred thousand	100 000th a/one hundred thousandth
/ə, wʌn ˌhʌndrəd 'θaʊznd/	/ə, wʌn ˌhʌndrəd 'θaʊznθ/
1 000 000 a/one million	1 000 000th a/one millionth
/ə, wʌn 'mɪlɪən/	/ə, wʌn 'mɪlɪənθ/

Some More Complex Numbers

101 a/one hundred and one /ə, wʌn ˌhʌndrəd n 'wʌn/
152 a/one hundred and fifty-two /ə, wʌn ˌhʌndrəd n ˌfɪftɪ 'tuː/
1 001 a/one thousand and one /ə, wʌn ˌθaʊznd ən 'wʌn/
2 325 two thousand, three hundred and twenty-five /ˌtuː ˌθaʊznd, θriː ˌhʌndrəd n ˌtwentɪ 'faɪv/
15 972 fifteen thousand, nine hundred and seventy-two /ˌfɪftiːn ˌθaʊznd, ˌnaɪn ˌhʌndrəd n ˌsevntɪ 'tuː/
234 753 two hundred and thirty-four thousand, seven hundred and fifty-three /ˌtuː ˌhʌndrəd n
  ˌθɜːtɪ fɔː ˌθaʊznd, ˌsevn ˌhʌndrəd n ˌfɪftɪ 'θriː/

		US	GB and other European countries
1 000 000 000	10^9	a/one billion	a/one thousand million(s)
		/ə, wʌn 'bɪlɪən/	/ə, wʌn 'θaʊznd 'mɪlɪən(z)/
1 000 000 000 000	10^{12}	a/one trillion	a/one billion
		/ə, wʌn 'trɪlɪən/	/ə, wʌn 'bɪlɪən/
1 000 000 000 000 000	10^{15}	a/one quadrillion	a/one thousand billion(s)
		/ə, wʌn kwɒ'drɪlɪən/	/ə, wʌn 'θaʊznd 'bɪlɪən(z)/
1 000 000 000 000 000 000	10^{18}	a/one quintillion	a/one trillion
		/ə, wʌn kwɪn'tɪlɪən/	/ə, wʌn 'trɪlɪən/

Vulgar Fractions

$\frac{1}{8}$ an/one eighth /ən, wʌn 'eɪtθ/
$\frac{1}{4}$ a/one quarter /ə, wʌn 'kwɔːtə(r)/
$\frac{1}{3}$ a/one third /ə, wʌn 'θɜːd/
$\frac{1}{2}$ a/one half /ə, wʌn 'hɑːf US: 'hæf/
$\frac{3}{4}$ three quarters /ˌθriː 'kwɔːtəz/

Decimal Fractions

0·125 (nought) point one two five /(ˌnɔːt) pɔɪnt ˌwʌn tuː 'faɪv/
0·25 (nought) point two five /(ˌnɔːt) pɔɪnt ˌtuː 'faɪv/
0·33 (ˌnought) point ˌthree 'three
0·5 (ˌnought) point 'five
0·75 (ˌnought) point ˌseven 'five

Notes. 1 In the spoken forms of vulgar fractions, the versions 'and a half/quarter/third' are preferred to 'and one half/quarter/third' whether the measurement is approximate or precise. With more obviously precise fractions like $\frac{1}{8}$, $\frac{1}{16}$, 'and one eighth/sixteenth' is normal. Complex fractions like 3/462, 20/83 are spoken as 'three over four-six-two; twenty over eighty-three', especially in mathematical expressions, e g 'twenty-two over seven' for 22/7.

2 When speaking ordinary numbers we can use 'zero', 'nought' or 'oh' /əʊ/ for the number 0; 'zero' is the most common US usage and the most technical or precise form, 'oh' is the least technical or precise. In using decimals, to say 'nought point five' for 0·5 is a more precise usage than 'point five'.

3 In most continental European countries a comma is used in place of the GB/US decimal point. Thus 6·014 is written 6,014 in France. A space is used to separate off the thousands in numbers larger than 9999, e g 10000 or 875380. GB/US usage can also have a comma in this place, e g 7,500,000. This comma is replaced by a point in continental European countries, e g 7.500.000. Thus 23,500·75 (GB/US) will be written 23.500,75 in France.

Collective Numbers

6	a half dozen/half a dozen	144	a/one gross /grəʊs/
12	a/one dozen (24 is two dozen *not* two dozens)		three score years and ten (Biblical) = 70 years,
20	a/one score		the traditional average life-span of man.

Roman		Arabic	Roman		Arabic	Roman	Arabic	Roman	Arabic
I	i	1	XVI	xvi	16	LX	60	DCC	700
II	ii	2	XVII	xvii	17	LXV	65	DCCC	800
III	iii	3	XVIII	xviii	18	LXX	70	CM	900
IV (IIII)	iv (iiii)	4	XIX	xix	19	LXXX	80	M	1000
V	v	5	XX	xx	20	XC	90	MC	1100
VI	vi	6	XXI	xxi	21	XCII	92	MCD	1400
VII	vii	7	XXV	xxv	25	XCV	95	MDC	1600
VIII	viii	8	XXIX	xxix	29	XCVIII	98	MDCLXVI	1666
IX	ix	9	XXX	xxx	30	IC	99	MDCCCLXXXVIII	
X	x	10	XXXI		31	C	100		1888
XI	xi	11	XXXIV		34	CC	200	MDCCCXCIX	1899
XII	xii	12	XXXIX		39	CCC	300	MCM	1900
XIII	xiii	13	XL		40	CD	400	MCMLXXVI	1976
XIV	xiv	14	L		50	D	500	MCMLXXXIV	1984
XV	xv	15	LV		55	DC	600	MM	2000

Weights and Measures[3]

The Metric System

The Metric System

METRIC	*length*	GB & US
10 millimetres (mm)	= 1 centimetre (cm)	= 0.3937 inches (in)
100 centimetres	= 1 metre (m)	= 39.37 inches or 1.094 yards (yd)
1000 metres	= 1 kilometre (km)	= 0.62137 miles or about ⅝ mile

	surface	
100 square metres (m²)	= 1 are (a)	= 0.0247 acres
100 ares	= 1 hectare (ha)	= 2.471 acres
100 hectares	= 1 square kilometre (km²)	= 0.386 square miles

	weight	
10 milligrams (mg)	= 1 centigram (cg)	= 0.1543 grains
100 centigrams	= 1 gram	= 15.4323 grains
1000 grams	= 1 kilogram (kg)	= 2.2046 pounds
1000 kilograms	= 1 tonne	= 19.684 cwt

	capacity	
1000 millilitres (ml)	= 1 litre (l)	= 1.75 pints (= 2.101 US pints)
10 litres	= 1 decalitre (dl)	= 2.1997 gallons (= 2.63 US gallons)

Avoirdupois Weight

GB & US		METRIC
	1 grain (gr)	= 0.0648 grams (g)
437½ grains	= 1 ounce (oz)	= 28.35 grams
16 drams (dr)	= 1 ounce	= 28.35 grams
16 ounces	= 1 pound (lb)	= 0.454 kilograms (kg)
14 pounds	= 1 stone	= 6.356 kilograms
2 stone	= 1 quarter	= 12.7 kilograms
4 quarters	= 1 hundredweight (cwt)	= 50.8 kilograms
112 pounds	= 1 cwt	= 50.8 kilograms
100 pounds	= 1 short cwt	= 45.4 kilograms
20 cwt	= 1 ton	= 1016.04 kilograms
2000 pounds	= 1 short ton	= 0.907 metric tons
2240 pounds	= 1 long ton	= 1.016 metric tons

Troy Weight

system of weights used in England for gold, silver and precious stones

GB & US		METRIC
24 grains	= 1 pennyweight (dwt)	= 1.555 grams
20 pennyweights	= 1 ounce	= 31.1 grams
12 ounces	= 1 pound (5760 grains)	= 0.373 kilograms

Apothecaries' Weight

used by pharmacists for mixing their medicines; they buy and sell drugs by Avoirdupois weight

GB & US		METRIC
20 grains	= 1 scruple	= 1.296 grams
3 scruples	= 1 dram	= 3.888 grams
8 drams	= 1 ounce	= 31.1035 grams
12 ounces	= 1 pound	= 373.24 grams

Linear Measure

GB & US		METRIC
	1 inch (in)	= 25.3995 millimetres (mm)
12 inches	= 1 foot (ft)	= 30.479 centimetres (cm)
3 feet	= 1 yard (yd)	= 0.9144 metres (m)
5½ yards	= 1 rod, pole, or perch	= 5.0292 metres
22 yards	= 1 chain (ch)	= 20.1168 metres
220 yards	= 1 furlong (fur)	= 201.168 metres
8 furlongs	= 1 mile	= 1.6093 kilometres (km)
1760 yards	= 1 mile	= 1.6093 kilometres
3 miles	= 1 league	= 4.8279 kilometres

Square Measure

GB & US		METRIC
	1 square inch	= 6.4516 sq centimetres
144 sq inches	= 1 sq foot	= 929.030 sq centimetres
9 sq feet	= 1 sq yard	= 0.836 sq metres
484 sq yards	= 1 sq chain	= 404.624 sq metres
4840 sq yards	= 1 acre	= 0.405 hectares
40 sq rods	= 1 rood	= 10.1168 ares
4 roods	= 1 acre	= 0.405 hectares
640 acres	= 1 sq mile	= 2.599 sq kilometres

Cubic Measure

GB & US		METRIC
	1 cubic inch	= 16.387 cu centimetres
1728 cu inches	= 1 cu foot	= 0.028 cu metres
27 cu feet	= 1 cu yard	= 0.765 cu metres

Surveyors' Measure

GB & US		METRIC
7.92 inches	= 1 link	= 20.1168 centimetres
100 links	= 1 chain	= 20.1168 metres
10 chains	= 1 furlong	= 201.168 metres
80 chains	= 1 mile	= 1.6093 kilometres
10 square chains	= 1 acre	= 0.405 hectares

Nautical Measure

used for measuring the depth and surface distance of seas, rivers, etc

GB & US		METRIC
6 feet	= 1 fathom	= 1.8288 metres
608 feet	= 1 cable	= 185.313 metres
6,080 feet	= sea (or nautical) mile (1.151 statute miles)	= 1.852 kilometres
3 sea miles	= 1 sea league	= 5.550 kilometres
60 sea miles	= 1 degree	
360 degrees	= 1 circle	

The speed of one sea mile per hour is called a *knot*

Liquid Measure of Capacity

	GB	US	METRIC
4 gills	= 1 pint (pt)	= 1.201 pints	= 0.5679 litres
2 pints	= 1 quart (qt)	= 1.201 quarts	= 1.1359 litres
4 quarts	= 1 gallon (gal)	= 1.201 gallons	= 4.546 litres

Apothecaries' Fluid Measure

used by pharmacists for measuring medicines

GB & US		METRIC
60 minims	= 1 fluid dram	= 3.552 millilitres
8 fluid drams	= 1 fluid ounce	= 2.841 centilitres
20 fluid ounces	= 1 pint	= 0.568 litres
8 pints	= 1 gallon	= 4.546 litres

Dry Measure of Capacity

	GB & US	METRIC GB	US
	1 gallon	= 4.5435 litres	4.404 liters
2 gallons	= 1 peck	= 9.0870 litres	8.810 liters
4 pecks	= 1 bushel	= 36.3477 litres	35.238 liters
8 bushels	= 1 quarter	= 290.7816 litres	281.904 liters

Circular or Angular Measure

60 seconds (″) = 1 minute (′)	90 degrees = 1 quadrant or right angle (L)
60 minutes = 1 degree (°)	360 degrees = 1 circle or circumference

The diameter of a circle	= the straight line passing through its centre
The radius of a circle	= ½ × the diameter
The circumference of a circle	= $^{22}/_7$ × the diameter

Irregular Verbs

to abide	abode, abided	abode, abided	(treu) bleiben
to awake	awoke	awaked, awoken	erwachen
to be	was/were	been	sein
to bear	bore	borne (born)	(er)tragen
to beat	beat	beaten	schlagen
to become	became	become	werden
to begin	began	begun	beginnen
to bend	bent	bent	beugen; biegen
to bereave	bereaved, bereft	bereaved, bereft	berauben
to beseech	besought	besought	anflehen
to bet	bet	bet	wetten
to bid	bade	bidden	heißen, gebieten
to bid	bid	bid	(als Preis) bieten
to bind	bound	bound	binden
to bite	bit	bit, bitten	beißen
to bleed	bled	bled	bluten
to blend	blended, blent	blended, blent	(ver)mischen
to blow	blew	blown	blasen
to break	broke	broken	brechen
to breed	bred	bred	züchten; erziehen
to bring	brought	brought	bringen
to broadcast	broadcast, -ed	broadcast, -ed	(im Rundfunk) senden
to build	built	built	bauen
to burn	burnt, burned	burnt, burned	(ver)brennen
to burst	burst	burst	bersten
to buy	bought	bought	kaufen
to cast	cast	cast	werfen
to catch	caught	caught	fangen
to chide	chided, chid	chided, chidden	schelten
to choose	chose	chosen	wählen
to cleave	clove, cleft	cloven, cleft	(zer)spalten
to cling	clung	clung	kleben; s. klammern
to clothe	clothed, clad	clothed, clad	(an-be-ein)kleiden
to come	came	come	kommen
to cost	cost	cost	kosten
to creep	crept	crept	kriechen
to cut	cut	cut	schneiden
to deal	dealt	dealt	(be-)handeln
to dig	dug	dug	graben
to do	did	done	tun
to draw	drew	drawn	ziehen; zeichnen
to dream	dreamt, dreamed	dreamt, dreamed	träumen
to drink	drank	drunk	trinken
to drive	drove	driven	treiben; fahren
to dwell	dwelt	dwelt	wohnen; verweilen
to eat	ate	eaten	essen
to fall	fell	fallen	fallen

to feed	fed	fed	füttern
to feel	felt	felt	fühlen
to fight	fought	fought	kämpfen
to find	found	found	finden
to flee	fled	fled	fliehen
to fling	flung	flung	schleudern
to fly	flew	flown	fliegen
to forbear	forbore	forborne	unterlassen
to forbid	forbade, forbad	forbidden	verbieten
to forecast	forecast	forecast	voraussagen
to foresee	foresaw	foreseen	voraussehen
to foretell	foretold	foretold	vorhersagen
to forget	forgot	forgotten	vergessen
to forgive	forgave	forgiven	vergeben
to forsake	forsook	forsaken	verlassen
to forswear	forswore	forsworn	(eidlich) bestreiten
to freeze	froze	frozen	(ge)frieren
to gainsay	gainsaid	gainsaid	bestreiten, leugnen
to get	got	got (gotten)	bekommen, werden
to gild	gilded, gilt	gilded	vergolden
to give	gave	given	geben
to go	went	gone	gehen
to grind	ground	ground	mahlen; schleifen; zer-reiben, -kleinern
to grow	grew	grown	wachsen; werden
to hang	hung	hung	(auf)hängen
to hang	hanged	hanged	an d. Galgen hängen
to have	had	had	haben
to hear	heard	heard	hören
to heave	heaved, hove	heaved, hove	auf-, hochheben
to hide	hid	hidden	(sich) verstecken
to hit	hit	hit	schlagen; treffen
to hold	held	held	halten
to hurt	hurt	hurt	(sich) verletzen
to keep	kept	kept	(be)halten
to kneel	knelt	knelt	knien
to knit	knitted, knit	knitted, knit	stricken
to know	knew	known	kennen, wissen
to lay	laid	laid	legen
to lead	led	led	leiten, führen
to lean	leant, leaned	leant, leaned	lehnen
to leap	leapt, leaped	leapt, leaped	springen
to learn	learnt, learned	learnt, learned	lernen
to leave	left	left	(ver)lassen
to lend	lent	lent	leihen
to let	let	let	lassen
to lie	lay	lain	liegen
to light	lit, lighted	lit, lighted	anzünden, erleuchten
to lose	lost	lost	verlieren

to make	made	made	machen
to mean	meant	meant	meinen; bedeuten
to meet	met	met	begegnen, treffen
to mow	mowed	mown; (US) mowed	mähen
to outdo	outdid	outdone	übertreffen
to outgo	outwent	outgone	übertreffen; -listen
to partake	partook	partaken	teilnehmen
to pay	paid	paid	(be)zahlen
to put	put	put	setzen, stellen, legen, stecken
to read	read	read	lesen
to ride	rode	ridden	reiten; fahren
to ring	rang	rung	läuten; erschallen
to rise	rose	risen	sich erheben
to run	ran	run	rennen
to saw	sawed	sawn	sägen
to say	said	said	sagen
to see	saw	seen	sehen
to seek	sought	sought	suchen
to sell	sold	sold	verkaufen
to send	sent	sent	senden, schicken
to set	set	set	setzen; stellen
to sew	sewed	sewn, sewed	nähen
to shake	shook	shaken	schütteln, zittern
to shed	shed	shed	vergießen
to shine	shone	shone	scheinen, glänzen
to shoot	shot	shot	schießen
to show	showed	shown	zeigen
to shrink	shrank, shrunk	shrunk, shrunken	schrumpfen; zurück-schrecken
to shut	shut	shut	zumachen
to sing	sang	sung	singen
to sink	sank	sunk, sunken	sinken, versenken
to sit	sat	sat	sitzen
to slay	slew	slain	erschlagen
to sleep	slept	slept	schlafen
to slide	slid	slid	gleiten
to sling	slung	slung	schleudern; hochziehen; umhängen
to slink	slunk	slunk	schleichen
to slit	slit	slit	(auf)schlitzen
to smell	smelt	smelt	riechen
to sow	sowed	sown, sowed	säen
to speak	spoke	spoken	sprechen
to speed	sped, speeded	sped, speeded	(dahin)eilen; beschleunigen
to spell	spelt, spelled	spelt, spelled	buchstabieren
to spend	spent	spent	ausgeben; verbringen
to spill	spilt, spilled	spilt, spilled	verschütten

to spin	spun, span	spun	drehen; spinnen
to spit	spat	spat	spucken
to split	split	split	spalten
to spoil	spoilt, spoiled	spoilt, spoiled	verderben; verziehen, verwöhnen
to spread	spread	spread	(sich) ausbreiten
to spring	sprang	sprung	springen
to stand	stood	stood	stehen
to steal	stole	stolen	stehlen
to stick	stuck	stuck	(an)kleben; stechen
to sting	stung	stung	stechen
to stink	stank, stunk	stunk	stinken
to strew	strewed	strewn, strewed	(aus-be)streuen
to stride	strode	stridden	schreiten
to strike	struck	struck, stricken	stoßen, schlagen
to strive	strove	striven	streben
to swear	swore	sworn	schwören; fluchen
to sweep	swept	swept	fegen
to swell	swelled	swollen, swelled	(an- auf)schwellen
to swim	swam	swum	schwimmen
to swing	swung	swung	schwingen
to take	took	taken	nehmen; bringen
to teach	taught	taught	lehren
to tear	tore	torn	zerreißen
to tell	told	told	erzählen; (es) sagen
to think	thought	thought	denken
to thrive	throve, thrived	thriven, thrived	gedeihen
to throw	threw	thrown	werfen
to thrust	thrust	thrust	stoßen
to tread	trod	trodden, trod	treten; schreiten
to understand	understood	understood	verstehen
to undo	undid	undone	aufmachen
to upset	upset	upset	umwerfen; bestürzen
to wake	woke, waked	woken, waked	(auf)wecken, -wachen
to wear	wore	worn	(an sich) tragen; abtragen
to weave	wove	woven	weben
to weep	wept	wept	weinen
to win	won	won	gewinnen
to wind	wound	wound	winden
to withdraw	withdrew	withdrawn	zurückziehen
to withhold	withheld	withheld	zurück-, abhalten
to withstand	withstood	withstood	widerstehen
to wring	wrung	wrung	auswringen
to write	wrote	written	schreiben

Quellenangaben

[1] Laboratory Manual for Chemistry. An Experimental Science (1963), W. H. Freeman and Comp., San Francisco, S. 134.

[2] Gilmore, G. N. (1982), A Complete 'O' Level Chemistry, Stanley Thornes LTD, S. 287, 288.

[3] Hornby, A. S. (1980), Oxford Advanced Learner's Dictionary of Current English, Oxford University Press, S. 1016, 1017.

[4] Dakin, J. (1975) Songs and Rhymes for the Teaching of English, Longman, London, S. 19.

[5] Alexeyev, V. Quantitative Analysis, Foreign Languages Publishing House, Moskau, S. 17.

[6] Kerstein, G. (1962), Entschleierung der Materie, Franckh'sche Verlagshandlung, Stuttgart, S. 35.

[7] Bild der Wissenschaft, Dezember 1982, S. 134.

[8] Time, November 8, 1982, S. 41, 42.

[9] Ceram, C. W., Götter, Gräber und Gelehrte, Rowohlt-Verlag, Hamburg.

[10] Fritz, M. (1977), Klipp und klar 100 × Umwelt, Bibliographisches Institut, Mannheim, Wien, Zürich.

[11] WTW-Konduktometer, aktuelles Gerät LF 530. Wissenschaftlich-Technische Werkstätten GmbH, 8120 Weilheim i. Ob., Trifthofstraße 57a

[12] English for Today (1975), Verlag Lambert Lensing, GmbH Dortmund, Hermann Schroedel Verlag KG, Hannover.

[13] Waldner Laboreinrichtungen, GmbH u. Co. D-7988 Wangen i. Allgäu P. O. Box 98

[14] IBN, Internationale Bodensee und Boot Nachrichten, 9/83.

[15] Basic English for Science (1978), Oxford University Press, Oxford.

[16] Earthwatch (1980), International Planned Parenthood Federation, London.

Literaturverzeichnis

Barclay, A. (1937), Pure Chemistry; A Brief Outline of Its History and Development; His Majesty's Stationery Office, London.

Davies, D., Locke, R. (1981), Investigating Chemistry, Heinemann Educational Books Ltd, London.

Jaffe, B. (1976), Crucibles: The Story of Chemistry, Dover Publications New York.

Vaclavik, J. (1978), Scientific and Technical English, Verlag der Fachvereine an den Schweizerischen Hochschulen und Techniken, Zürich.

Bandtock, J., Hanson, P. (1975), Success in Chemistry, John Murray Ltd. London.